INFORMATION WARFARE

Issues in Twenty-First Century Warfare

SERIES EDITORS

Capt. Roger W. Barnett, USN (Ret.), Ph.D.
Naval War College

Stephen Cimbala, Ph.D.
Penn State University

ADVISORY BOARD

Col. C. Kenneth Allard, USA (Ret.), Ph.D.
Potomac Strategies International, LLC

Capt. John P. Cann, USNR (Ret.), Ph.D.
U.S. Marine Corps Command and Staff College

Colin Gray, D. Phil.
Centre for Strategic Studies, University of Reading

Benjamin S. Lambeth, Ph.D.
RAND

Mark D. Mandeles, Ph.D.
The J. de Bloch Group

Michael F. Pavkovič, Ph.D.
Hawaii Pacific University

Capt. Peter M. Swartz, USN (Ret.)
CNA Corporation

INFORMATION WARFARE

Separating Hype from Reality

Edited by

LEIGH ARMISTEAD

Potomac Books, Inc.
Washington, D.C.

Library of Congress Cataloging-in-Publication Data

Information warfare : separating hype from reality / Edited by Leigh Armistead.
    p. cm.
Includes bibliographical references.
ISBN-13: 978-1-59797-057-0 (hardcover : alk. paper)
ISBN-10: 1-59797-057-3 (hardcover : alk. paper)
ISBN-13: 978-1-59797-058-7 (pbk. : alk. paper)
ISBN-10: 1-59797-058-1 (pbk. : alk. paper)
1. Information warfare. I. Armistead, Edwin Leigh.

U163.I54 2007
355.3'43—dc22
                                            2006027135
Hardcover ISBN-10: 1-59797-057-3
Hardcover ISBN-13: 978-1-59797-057-0

Softcover ISBN-10: 1-59797-058-1
Softcover ISBN-13: 978-1-59797-058-7

(alk. paper)

Printed in the United States of America on acid-free paper that meets the American National Standards Institute Z39-48 Standard.

Potomac Books, Inc.
22841 Quicksilver Drive
Dulles, Virginia 20166

First Edition

10 9 8 7 6 5 4 3 2 1

Contents

CHAPTER 1

Introduction: "Brother, Can You Spare Me a DIME?"

Dr Dan Kuehl, Information Operations Concentration, National Defense University

One of the central objectives of this book is to thoroughly ground Informa tion Operations (IO) in the real world and to concentrate on IO's actual challenges, capabilities, and accomplishments. Earlier claims by IO en- thusiasts of "bloodless wars" or teenage hackers plunging North America into a new "dark age" of constant collapses of the electric grid raised awareness of new threats and capabilities in this arena, but many also went beyond the believability limit, and thus may have hampered acceptance and understanding of those threats and capabilities. This book helps to clarify these misunderstandings.

Our educational venues are especially in need of a corrective, and this is one of this volume's key objectives, particularly at the strategic level. The Department of Defense's (DOD) schools and systems for training and education does an ex- cellent job of producing information "warriors," those who are expert in employ- ing the various Information Operations core competencies—Psychological Operations, Military Deception, Operational Security, Electronic Warfare, and Computer Network Operations. This is where the armed forces so effectively per- form their mission of "organizing, training, and equipping" military forces, and there are myriad courses and programs across the DoD that prepare personnel in these competencies. We are doing an increasingly effective job of preparing infor- mation "planners," those who can coordinate and integrate the several IO compe- tencies into an IO plan, and then integrate that into a theater operations plan. This is the new paradigm for joint warfare: not the integration of the Services, but rather the integration of operations across battlespaces—air, land, sea, space, and information. Where we need the greatest improvement is in the development of information "strategists," those who are able to coordinate and exploit the contri- bution of the information component of power and the synergies it offers to the other elements of national power. This is the source of this brief introduction's subtitle, because our military education programs and curricula need to more effectively present the information component of national power—the "I" in the acronym that represents the Diplomatic-Information-Military-Economic elements of national power—especially how it shapes and impacts the ways the other elements are developed and employed. This book presents the reality of this element of power and its impact on national security.

A second strategic reality resides in the environment through which this element of power is employed. The cover of *Information Operations: Warfare and the Hard Reality of Soft Power*, the previous book in this series, featured a photograph of a satellite dish antenna, which illustrates simultaneously an opportunity and a problem. The opportunity comes from the spectacular and seemingly limitless advances in the technologies we use to store, shape, and share information. The problem is that that far too often anything related to Information Operations is seen through the single-focus lens of technology, when in fact the issue is much deeper and more complex. The new joint doctrine for IO, however, contains an approach that is based on perspectives we have been teaching at the National Defense University for several years and which mark a significant improvement over previous constructs. The new approach characterizes the information environment as a trinity of three distinct yet interrelated and integrated elements. The physical dimension constitutes the interconnectivity of information technologies and is the stuff that we see and use everyday: wires, networks, phones, computers, etc. The information dimension is the content carried by those interconnected systems such as TV broadcasts, radio programs, databases, and phone calls. The cognitive dimension is the third and most important of them, because this is where the content that is delivered by the connectivity impacts human beings and how we think, decide, and act. This approach may help connect the theory to the reality, because we see evidences of each of these three domains and their interrelationships every day.

The connectivity is perhaps the most visible, because we use and touch multiple means in our daily work and life. We work in offices and headquarters that have telephones, faxes, computers, radios, and TV that are connected via coaxial cables, fiber optics, satellite dishes, and even wi-fi networks. The modern Stryker combat vehicle has more information connectivity than Admiral Jellicoe had on his command battleship at the Battle of Jutland. The content it carries is all-pervasive because we are awash in a sea of it: we see the latest news from overseas, we listen to the all-important traffic updates while we drive, our medical test results are faxed or even emailed to the doctor while we are being treated, and our weaponry is constantly updated for precision as to its location as well as its target. But while the least visible domain—the cognitive—may be the most difficult to directly tie to results, it is where everything comes together, and examples are legion across all of the realms of human activity, from global economics to diplomatic influence to military strategy, demonstrating how our perception of reality has shaped our decisions and actions. While we often have difficulty establishing a direct cause-effect relationship between information actions we take and resultant behaviors and decisions, the better we can plan and execute information operations in the physical and content domains, the more effective we will be in creating results in the cognitive domain.

A third set of IO's realities can be seen in its use across the DoD and indeed the larger national security community. Each of IO's five core competencies saw multiple examples of effective employment throughout the past decade, from the Balkans to Iraq and Afghanistan and beyond. The several related and supporting activities outlined in the Secretary of Defense's (SecDef) "IO Roadmap" also saw multiple and successful uses, both as adjuncts to combat operations and as critical aspects of humanitarian or peacekeeping operations, from Kosovo to Kabul and Indonesia. But the reality extends well beyond the DoD, and extends from one of America's oldest institutions, the Department of State's Foreign Service, to the newest, the Department of Homeland Security's Cybersecurity Division. Internal government tests have demonstrated the vulnerabilities of computer-controlled systems, and terrorist laptops found in Afghan caves reportedly contained detailed information on the control systems of nuclear powerplants. Why? Probably not to complete their online version of "Terrorism 101." Numerous studies and reports over the past few years have made a convincing argument that in the struggle to sway global opinion and influence other populations, we are not winning and may be losing. The most senior officials in the U.S. government, including the Vice President, Secretary of Defense and Secretary of State, have echoed the realization that we have not waged this struggle effectively, nor used all of our tremendous informational capabilities that span so many media.

Thus it is crucial that we set aside the hype and conjecture concerning IO, because the reality is far more powerful and comprehensive. Information, as an element of power and a field for military operations and warfare, is critical to our military capability and future national security strategy. Our success on the battlefield, our economic and societal health and well-being, and our ability to win the global contest for influence all may depend on how smartly and effectively we wage Information Operations. This book takes us several steps along that path.

CHAPTER 2

Information Operations: The Policy and Organizational Evaluation

Dr Dan Kuehl, Professor of Information Operations,
Information Resources Management College,
National Defense University
Leigh Armistead, Edith Cowan University and
*Strategic IA Manager, Honeywell**

In an age when terrorists move information at the speed of an email, money at the speed of a wire transfer, and people at the speed of a commercial jetliner, the Defense Department (DoD) is bogged down in the micromanagement and bureaucratic processes of the industrial age, not the information age. Some of our difficulties are self-imposed, while others are the result of law and regulation. Together they have created a culture that too often stifles innovation, and results in the United States fighting the first wars of the twenty-first century with a Defense Department that was fashioned to meet the challenges of the mid-twentieth century. We have an industrial age organization, yet we are living in an information age world, where new threats emerge suddenly, often without warning, to surprise us. We cannot afford not to change and rapidly, if we hope to live in that world.[1]

This comment by former Secretary of Defense Donald Rumsfeld emphasizes the dichotomy that exists in the federal government. In this book, the authors attempt to explain the only the most recent policy and organizational changes, which have occurred within Information Operations (IO) over the last few years. For a more comprehensive and historical review of the early United States IO structure, please see the preceding publication *Information Operations: Warfare and Hard Reality of Soft Power*.

This chapter will trace the policy and organizational shifts within three distinct areas:

- Information Assurance (IA) and its related field, Critical Infrastructure Protection
- The military aspects of IO
- Strategic Communication (SC) and its related field, Public Diplomacy

For the goal of this book is to move away from some of the more sensational aspects of IO that have been promoted over the last 10 years and to show the "normalization" of IO as what the Secretary of Defense termed a "core military competency" that stands alongside of and is simultaneously supportive and yet distinct from the other realms of military activity, the land, air, sea, and special operations.[2]

Policy Changes: Defensive IO

In the five years since the events of September 11, much has transpired in the United States and around the world with respect to IO and its various supporting and related activities. Policy and organizational changes that were only dreamed of before this terrorist attack have now been realized. The most prominent example of this is the creation of a new cabinet-level agency, the Department of Homeland Security (DHS), which has assumed responsibility for oversight and management of the government's efforts and programs in Information Assurance and Critical Infrastructure Protection (IA/CIP). Began with the 1998 issuance of Presidential Decision Directive (PDD) 63 by the Clinton Administration, the events of 9/11 effected this area greatly, and the Bush Administration has followed this initial effort with several policy and organizational changes of its own. Although the *National Strategy to Secure Cyberspace* was issued after the terrorist attacks, the strategy was written and coordinated before that date, and reflected the efforts of the Bush Administration's then-advisor for infrastructure protection, Richard Clarke, who had also had the major hand in the Clinton Administration's efforts in this area. In 2003 the Bush Administration also issued two further strategies, the *National Strategy for the Physical Protection of Critical Infrastructures and Key Assets*, and the Homeland Security Presidential Directive 7, *Critical Infrastructure Identification, Prioritization, and Protection*. In 2006 these efforts were extended with the *National Infrastructure Protection Plan*. All of these strategy and guidance documents reflected the same basic philosophy as did PDD-63, namely that the task of IA and CIP at the national level was too difficult a task for unilateral government or business-sector solutions and thus required a partnership between all parties: owners, users, and the national security apparatus.

The Defense Department also recognized the importance of this issue long ago. Indeed the DOD was one of the instigators of national-level studies that began in the early 1990s, and within the Joint Staff, which is responsible for DoD-wide communications, the J-6 directorate has been one of the central players in this area. In early 2006, the J-6 created a new office, the J-6X, and assigned it the responsibility of developing a *National Military Strategy to Secure Cyberspace*, the name being chosen to obviously parallel the 2001 *National Strategy to Secure*

Cyberspace. Although this effort was unable to meet its initial and overly-ambitious timeline of 120 days from start to finish, by the end of 2006, the new *National Military Strategy for Cyberspace Operations* had been signed by the Chairman of the Joint Chiefs of Staff. One of the reasons for the change in title was the belief that the original was too defensive in nature, and what was really needed was a policy that was more proactive and inclusive.[3]

This small change is part of a broader discussion by the DoD on the alignment of IO into offensive and defensive capabilities that match better to their functional organizations. For if International Public Information (Clinton Administration) or Strategic Communication (Bush Administration) and IO are normally considered the "offensive" aspects of this warfare area, than IA with its related functions of CIP and Computer Network Defense (CND) are more in the defensive realm. In fact the forerunners of IA in the form of Information Security (INFOSEC) and Computer Security (COMPUSEC) have long and distinguished histories within the defense bureaucracy. A good example of this regards a portion of IA that centers around computer security assessments plus the certification and accreditation (C&A) process. The current methodology for information assurance is known as the *Department of Defense Information Technology Security Certification and Accreditation Process (DITSCAP)*, which has been in existence for nine years and it is being replaced by a new C&A policy entitled the *Department of Defense Information Assurance Certification and Accreditation Process (DIACAP).*[4] What this new process does is to force the program managers to evaluate their system from a confidentiality, integrity and availability standpoint on the value of the information protected. To do this, the program managers must determine the confidentiality, robustness and mission assurance category of their architecture by discussing and analyzing the system with key personnel, such as the user representatives, system administrators, information system security managers and certification agent. This doctrine was a concerted attempt by the Office of the Secretary of Defense (OSD) to lay out a new methodology for ensuring the security of its networks and applications, by standardizing the process through well-recognized IA controls. This is important because this new policy tightens the protection of the government and DoD by enforcing standards across the enterprise.

In addition to the directives tasking IA policy, the DoD has also developed a series of standards to map skill sets of different levels of IA professionals. As Dr Corey Schou, Director of the National IA Training and Education Center (NIATEC) relates, the security community did not begin to recognize the need for improved education and training until 1986.[5] To encourage academic participation, they established training standards and a defined set of knowledge skills and attributes.[6] This early work at NIATEC to develop a taxonomy, led to several industry professional standards, National Institute of Standards (NIST) publication 800-16, and the National Security Telecommunications and Information Systems Security Committee (NSTISSC) series of publications. These standards, developed by the NSA, are now widely recognized throughout the DoD and interagency as the de facto baseline of tasks for IA across the federal bureaucracy, and the Committee on

INFORMATION WARFARE

National Security Systems (CNSS) series has become widely used in academia, through NSA-sponsored IA programs and curriculum. For under Executive Order 13231 (October 16, 2001), *Critical Infrastructure Protection in the Information Age*, which will be alluded to in the next section as well, the Bush administration redesignated the NSTISSI as the CNSS. This group provides a forum for the discussion of policy issues, sets national policy, and promulgates direction, operational procedures, and guidance for the security of national security systems through the CNSS Issuance System as shown in the NSTISSI and CNSS documents 4011–4015.

Likewise, from a CIP standpoint, there has been an equally prodigious output of directives and memorandum from the Clinton and Bush Administrations including as an example, three Executive Orders (13010, July 15, 1996), (13064, October 11, 1997), and (13231, October 16, 2001), a Homeland Security Presidential Directive (HSPD) 7, (December 17, 2003), and three Government Accountability Office (GAO) Reports (March, April, and May 2004), all of which were produced over an eight year period to support this area of IA. What this effort did was to bring together the disparate elements of the business and governmental interests, to move forward CIP as a useful component of IO. However, because most of the infrastructure portion of CIP is predominantly owned and operated as a function of the commercial sector, progress has been uneven, with some segments, notably banking and finance, advancing more rapidly than others. This disparate focus is especially noted in the three GAO reports that highlight deficiencies in not only the efforts of the business sector but the federal government as well.

There have also been directives on CND such as *The National Strategy to Secure Cyberspace* (2003), which ties into CIP as part of a larger effort to protect America. An implementing component of *The National Strategy for Homeland Security* (2002) and complemented by a *National Strategy for the Physical Protection of Critical Infrastructures and Key Assets* (2003), all of these documents were developed to allow the American public and commercial industries to secure the portions of cyberspace that they own, operate, control, or with which they interact. Once again, these documents reiterate one of the key lessons of this process, namely that IO does not have to be a top-down effort, because power has been shifted to the masses as part of the information age, but the protection of America must now be disseminated as well. Citizens of the United States are very accustomed to having the military or armed forces act as their protector against adversaries, but in the information age that is not always possible or practical. Securing the population is a difficult strategic challenge that requires coordinated and focused effort from our entire society, the federal government, state and local governments, the private sector, and the American people. That is what is different about this new era and what must be accepted in order to truly understand the power inherent in information.

The final new policy and organizational initiative from a defensive IO perspective, has actually been the creation and development of a Department of Homeland Security. During the fall of 2000 and the spring of 2001, a 14-member bipartisan commission headed by former Senators Gary Hart (D-CO) and Warren Rudman (R-NH) released a three part series on the new threats to national security. Entitled the U.S. Commission on National Security/Twenty-First Century, their initial report *Road Map for National Security: Imperative for Change*, attempted to summarize, based upon the changing environment, the new threats to the United States, especially with respect to information.[7] These reports proposed radical changes in the structures and baseline processes of the governmental apparatus to ensure that America did not lose its global influence or leadership role. In an eerie coincidence (or perhaps not), the recommendations provided by this group provided much of the foundation for the changes that occurred after the attacks of September 11th. While initially scoffed at by academia and the federal bureaucracy, the suggestions of this commission on national security in fact foreshadowed much of the changes that have occurred over the last five years. Equally as disturbing with regard to threats to national security and the role of information was a series of comments made by then CIA Director George J. Tenant before the U.S. Senate select Committee on Intelligence on 7 February 2001. In this testimony, Tenant stated that

> The threat from terrorism is real, it is immediate, and it is evolving. . . . Terrorists are also becoming more operationally adept and more technically sophisticated . . . for example, as we have increased security around government and military facilities, terrorists are seeking out "softer" targets that provide opportunities for mass casualties.

While the true role of the DHS has still not been totally sorted, its organizational structure continues to evolve, and it remains to be seen how the interagency organizations ultimately adapt to these policy changes with respect to IO.

Policy Changes: Offensive IO

Undoubtedly the most significant policy change that impacts offensive IO, from an American standpoint, was the publication of the IO Road Map by the DoD in October 2003. This directive proposes a way ahead for the United States military forces with regard to the future of IO. The 2001 Quadrennial Defense Review (QDR) identified IO as one of six critical goals supporting DoD transformation, and it set forth the objective of making IO a "core capability" for future US forces. The IO Roadmap identified three critical areas in which US capabilities must be improved. The first was an improved ability to "fight the net", and stemmed from the realization that in an era of "network centric warfare, protecting the networks on which the DoD depends is an essential to US military capability. Second was the need to "improve PSYOP" (Psychological Operations), to make it more integrated with and supportive of national level themes and objectives, and to enhance US ability to impact adversary decision-making. Finally, the third crucial area was the need for the US forces to conduct offensive operations in/via the electromagnetic spectrum must be improved, to include both Computer Network Attack (CNA) and Electronic Warfare (EW) capabilities.[8]

The Roadmap recommended a series of actions to improve overall offensive IO capabilities.[9] The first was to develop a common understanding of IO, and it offered a new definition of IO that would eventually be issued to the joint world doctrinally via a revised *Joint Doctrine Pub 3-13*, issued in February 2006, and a revised *DoD Directive 3600*, issued in August 2006. The IO Roadmap stressed the need to both consolidate oversight and advocacy for IO, while simultaneously delegating capabilities to the Combatant Commanders, and to do this the US Strategic Command's IO role was expanded and strengthened, to the point where STRATCOM became in effect "the IO command". The need to create a core of trained and educated IO personnel, and the requirement to improve the ability to analyze IO operations and effects, were both cited in the IO Roadmap's recommendations. There were also recommendations for improvement of each of the five "core competencies" of IO as defined by the Roadmap – namely Computer Network Operations (CNO) (which includes attack, defense and exploitation), EW, Military Deception (MilDec), Operations Security (OPSEC), and PSYOP. The need to clarify the "lanes in the road" between PSYOP, Public Affairs (PA), and Public Diplomacy was also emphasized. Finally, IO's place in the budget process needed increased transparency, to clarify what resources IO actually had and what would be needed to provide a stronger, more robust and more comprehensive set of capabilities. The full IO Roadmap laid out 57 specific recommendations designed to develop specific elements of the overall recommendations discussed above.

The new definition of IO published in the Roadmap was very much centered on the military aspects of information, and was almost a verbatim return to that contained in the early 1990s doctrine for Command-and-Control Warfare (C2W), defining IO as "The integrated employment of the core capabilities of Electronic Warfare, Computer Network Operations, Psychological Operations, Military Deception, and Operations Security, in concert with specified supporting and related capabilities, to influence, disrupt, corrupt or usurp adversarial human and automated decision-making while protecting our own."[10] This new definition was a significant narrowing of IO's scope downward from what had been laid out in the 1998 *Joint Doctrine Pub 3-13*, which defined IO as "actions taken to affect adversary information and information systems while protecting our own." That earlier approach was much broader and more inclusive of other federal IO activities, and tended to focus on effects rather than means. It was also more difficult to resource. The military services are responsible for "organizing, training, and equipping" forces, and they complained that nothing in the original 1998 definition could be directly tied to military programs. The new IO Roadmap definition, on the other hand, could also be immediately tied to several long-standing and well integrated force programs. The new definition also included several controversial elements, most of which were related to the word "influence". The "lanes in the road" issue contained in the recommendations was the most controversial element, because it brought together into one discussion several activities and communities that traditionally have viewed each other with great suspicion. The links and relationships between Public Affairs (whether DoD or State), PSYOP, Military Deception, and Public Diplomacy (at State, which viewed the new IO term "defense support to public diplomacy" with skepticism) are undeniable in a theoretical sense, but in the "real world" of the federal government where turf battles, organizational cultures, and concerns over roles and responsibilities, all intermix to create an environment that often does not embrace change.

The IO Roadmap's definition was actively formalized across the Joint Force with the release of the newly-revised *Joint Doctrine Pub 3-13* in February, 2006. The old JP 3-13 had been in effect for more than seven years, during which much had changed in the IO environment. While the old doctrine had perhaps emphasized organizational measures, the new one makes several conceptual advancements. It described the information environment as a synergistic interaction of three dimensions: the physical, with the infrastructures and links of information networks; the informational, representing the actual material being carried by the physical networks; and the cognitive, of the perceptual element, where the human mind applies meaning to the information and which was described as the "most important" of the three. It also removes the term Information Warfare (IW) from the official lexicon, and while most of the rest of the world still uses IW as the most descriptive and commonly understood term for this, the DoD on the other hand has officially dropped it. The new JP 3-13 explicitly links IO to the Defense Department efforts to "transform" itself, and it emphasizes the importance of IO's multinational and coalition elements. The role of US STRATCOM as the chief advocate and proponent for IO is emphasized, and its mission of coordinating IO across geographic areas of responsibility, such as between combatant commands in Europe and Asia, and across functional boundaries, is described in greater detail than before. The relationship between Strategic Communication (SC) and IO is also emphasized, and it provides a definition, albeit perhaps a misguided one and misleading one for information superiority.[11]

In addition, the military services have also either published or revised their doctrines for IO in the last few years. The Marine Corps published *Marine Corps Warfighting Publication 3040.4, Marine Air-Ground Task Force Information Operations* in July 2003, the Army published a new *Field Manual (FM) 3-13 Information Operations* in November 2003, while the USAF published a new *Air Force Doctrine Document (AFDD) 2-5 Information Operations* in January 2005.[12] All of these reflected their individual Service's perspectives on warfare and IO, and not surprisingly viewed IO through the lenses of air, land or naval warfare. The final policy action to be discussed came in late 2006 with the USAF's "claiming" of cyberspace as one of its three core operational environments. While some saw this as nothing more than a turf grab for new missions and resources, in truth the Air Force had stated for more than a decade that it operated across three physical environments: air, outer space, and cyberspace. In December 2005, the USAF Chief of Staff, General Michael T. Moseley, and Secretary of the Air Force Michael

Wynne, issued a new USAF mission statement declaring that cyberspace was a core mission area for the Air Force, and they followed this policy statement in late summer 2006 with actions to create an Air Force major command for cyberspace operations that will stand alongside of Air Combat Command and Air Force Space Command.[13]

Yet the most radical change with regard to offensive IO policy changes have occurred in the realm of SC, Public Diplomacy (PD), International Public Information (IPI), Perception Management (PM) and PSYOP. A logical place to start this discussion about what changes have occurred in the post-9/11 world will more often than not involve the interagency cooperation and coordination efforts. This is because while no National Security Presidential Directive (NSPD) has been released on SC or IO, a significant number of new strategic guidance directives have been published, beginning with the new *National Military Strategy* published in 2005, and the new *National Military Strategy on Cyberspace Operations*, all of which require significant coordination across the different federal agencies. Likewise other key National Strategies addressing on Cyber Security, Homeland Security, and CIP have also been approved, which help to give an overarching framework to IO. Also as will be addressed later, while the two Policy Coordination Committees created by NSPD 1 remain in existence, in April 2006 a new PCC for Public Diplomacy and Strategic Communication was created, chaired by the Under Secretary of State for Public Diplomacy and Public Affairs, Ms. Karen Hughes.[14]

From the perspective of PM and SC policy for the debate over how to effectively communicate with and influence foreign populations and governments could be said to have begun with the publication of the Clinton era PDD 56, *Managing Complex Contingency Operations* (May 1997). Written after the debacles in Somalia, Haiti, and Rwanda, this directive was developed to integrate political, military, humanitarian, economic, and other dimensions of USG planning for complex contingencies, which included the informational aspects. Widely lauded at the time, subsequent studies and commentary reflect that in fact, little was changed by this PDD within the beltway.[15] Unfortunately if one examines the attempts to develop more recent and overarching IO doctrine with respect to interagency aspects of the "softer" side of IO, (PSYOP, IPI, PD or SC), these efforts have also been less than successful. Even before the events of 11 September 2001, there had been efforts by the White House to update and rewrite a new NSPD to focus on influence at the strategic level. A Defense Science Board (DSB) report on *Managed Information Dissemination (DSB-MID)* was released in October 2001. Written by public diplomacy professionals and led by its Chairman Vince Vitto during the transition period between the Clinton and Bush administrations, it laid the groundwork for the 2004 DSB Report on SC, but because it did not come from the Executive Branch, much of it's effectiveness was lost. In addition a new NSC Policy directive on SC, which was to rely on the three earlier National Security Council (NSC) directives (NSC 4 (1947), NSDD 130 (1984), and PDD 68 (1999)), was supposed to be issued in 2002. However that did not occur, for a variety of reasons, mostly political.[16] Some of this may have been due to the debacle concerning the Office of Strategic Influence (OSI) in February 2002, which in effect hamstrung the Bush administration in its attempts to develop a strategic communication effort. Thus the ultimate failure of the executive branch to promulgate a strategic policy in this area of IO probably occurred more as a case of general inertia and political unwillingness, than any other factor.

Yet these failures are not totally representative of all efforts on the offensive aspects of IO. In fact a number of significant changes have occurred within the U. S. government with respect to broader policy area of PD. For example, the term Strategic Influence has disappeared in lieu of the term SC, and since April 2002, the DoD has regrouped and pressed on to conduct strategic influence operations under this new name.[17] However even then, progress has been somewhat slow, and in many cases very sporadic. For example, the current structure of the new PCC for Public Diplomacy and Strategic Communication within the National Security Council (NSC) constitutes an attempt by the Bush administration to develop a SC capability. This new PCC, under the chairmanship of the Under Secretary of State for Public Diplomacy and Public Affairs, Karen Hughes, will attempt to oversee an overarching U. S. government strategy for Strategic Communications. This is very ironic, because it was only in late 1999 that the United States Information Agency (USIA) was dismantled, and its functions shifted under the greater umbrella of the State Department (DoS) in what many saw as a hostile takeover. In fact, in three successive years (2002, 2003, and 2004) Representative Henry Hyde (R-NY) proposed the reconstitution of the USIA, in a number of legislation attempts such as *Information Protection Act of 2002* - HR 3969.[18] But none of this legislation was successful, with these efforts mirroring the 2001 DSB-MIB, yet to date the State Department has not agreed with this concept, and therefore in the long run, nothing has ultimately come of these efforts to rebuild a USIA-like capability. Therefore the demise of the USIA may have contributed more to the need for a new office in the NSC, new presidential guidance, and the creation of the new PCC for PD/SC, than any other action to date.

Other changes are also still occurring with respect to the relationship between IO and the larger issue of strategic communication and influence. As the military conflicts in Afghanistan and Iraq have continued, more recommendations continue to come from various independent and quasi-government efforts regarding the need for a greater perception management capability by the United States government, as shown by some of the more recent efforts:

■   Building America's Public Diplomacy Through a Reformed Structure and Additional Resources, U.S. Advisory Commission on Public Diplomacy (2002),
■   U.S. Public Diplomacy, U.S. General Accounting Office (2003),
■   Finding America's Voice: A Strategy for Reinvigorating U.S. Public Diplomacy, Council on Foreign Relations (2003),

- Strengthening U.S.-Muslim Communications, Center for the Study of the Presidency (2003),
- How to Reinvigorate U.S. Public Diplomacy, Heritage Foundation (2003),
- The Youth Factor: The New Demographics of the Middle East and the Implications for U.S. Policy, The Brookings Institution (2003),
- Changing Minds, Winning Peace: A New Strategic Direction for U.S. Public Diplomacy in the Arab and Muslim World, U.S. Advisory Commission on Public Diplomacy (2003).

But even with the release in September 2004 of a new DSB Task Force *Report on Strategic Communications*, there still has been no official government action in this area. This report is a follow-up to the earlier 2001 study by the DSB, as mentioned earlier. Many critics felt the first study was overshadowed by the tragic events of 11 September and the opening campaign of Operation Enduring Freedom in Afghanistan, so a primary duty of this new *Report on Strategic Communications* was to not only look at the changes that had occurred since the original report, but also to reflect on the prior publication to see if its recommendations were still valid. The authors of this chapter believe that it is, and have included the opening statement, which is so powerful that it is worth repeating verbatim:

This Task Force concludes that U.S. strategic communication must be transformed. America's negative image in world opinion and diminished ability to persuade are consequences of factors other than failure to implement communications strategies. Interests collide. Leadership counts. Policies matter. Mistakes dismay our friends and provide enemies with unintentional assistance. Strategic communication is not the problem, but it is a problem.[19]

The report went on to cite seven key factors for success with regards to strategic communications by the United States. All of these areas were important, but without an administration and federal bureaucracy that understood the problem, gave strong leadership, and encouraged strong Government-Private Sector partnerships, this team saw little chance of success, notwithstanding its recommendations, which are laid out below:

- Issue a National Policy Security Directive on Strategic Communications from the NSC
- Establish a permanent strategic communication structure within the NSC to include a Deputy National Security Advisor and a SC Committee
- Create an independent, non-profit, and non-partisan Center for SC to support the NSC
- Redefine the role and responsibility of the Under Secretary of State for Public Diplomacy and Public Affairs to be both policy advisor and manager for PD

- The Public Diplomacy Office directors in the Department of State should be at the level of Deputy Assistant Secretary or Senior Adviser to the Assistant Secretary,
- The Under Secretary of Defense for Policy should act as the DoD focal point for SC
- The Under Secretary of Defense for Policy and the JCS ensure that all military plans and operations have appropriate strategic communication components[20]

What is very interesting from an academic standpoint is that many of the personnel interviewed for this DSB have also participated in a number of studies with the authors of this book.[21] Thus we feel the recommendations of this report mirror much of the contributors' research since 1999. In addition, all the key interviewees of the DSB worked at one time or are still associated with the public diplomacy, strategic communication, or international public information community, which validated their findings.[22] Finally, the appointment of Karen Hughes by the Bush administration in 2005, to the Under Secretary of State for Public Diplomacy and Public Affairs, was perhaps not only an attempt to answer its critics as well as these myriad of think tank reports but also to strengthen this aspect of US National Security policy.

This idea is crucial because as the events of September 11 indicate, military, political, or economic power is often ineffective in dealing with these new kinds of threats to the national security of the United States. These attacks were a blow to the American public and its perception of the government, and the fear produced by the terrorist acts can only be defeated by using a comprehensive plan in which information is a key element, or as John Arquilla and David Ronfeldt argued, networks fighting networks.[23] Both Operation Enduring Freedom and Operation Iraqi Freedom represent campaigns fought about perceptions, and the side that will ultimately emerge as the victor is the one that can best shape and influence the minds of not only their adversary, but their allies and even neutrals and uncommitted parties as well. The changes are truly revolutionary and describe a profound shift in the nature of power. Unfortunately, this transformation has not been translated from a strategic concept to tactical actions.[24]

The end of 2006 finally saw the emergence of two additional pieces of strategic guidance and policy, one from the DoD and one at the interagency level. In September 2006, the DoD released the Quadrennial Defense Review Execution Roadmap for Strategic Communication, which briefly summarized the problem facing the DOD and laid out 55 tasks intended to remedy those problems. The Roadmap began by defining Strategic Communication as "Focused US Government processes and efforts to understand and engage key audiences to create, strengthen or preserve conditions favorable to advance national interests and objectives through the use of coordinated information, themes, plans, programs, and actions synchronized with other elements of national power." This approach was significantly better than previous ones that emphasized the "transmission of themes

*a*nd messages". The new approach recognized that if one hopes to have any likelihood of positively influencing an audience, the first step must be listening to and
understanding that audience, and thus hopefully avoiding the widespread (and
sometimes accurate) global perception that the US is so busy talking that it can't
afford the time and effort to listen. The Roadmap also stated that the "U.S. military is not "sufficiently organized, trained, or equipped" to engage in full-spec-
trum strategic communication, and that "changes in the global information
environment" require a more coordinated and integrated effort. It emphasized the
importance of "credibility and trust", and noted that that all elements of the US
Government share the responsibility for this. For not only is effective SC a government-wide responsibility, the DoD is by no means the senior player in this
effort, and in fact it must support the efforts of the State Department to integrate
these efforts. Within the DoD, however, several key capabilities require improvement, most of which fall within the umbrella of IO in some way, including PA,
PSYOPS and Defense Support to PD. The DoD also defined three key objectives
in this Roadmap that if met would significantly improve its ability to conduct
effective Strategic Communication. First, the Defense Department needed to in-
stitutionalize a process through which goals and objectives in this issue area which
could be embedded within the development and execution of plans across all operational levels. Next, the doctrine needed to be developed to clearly define the
roles, responsibilities and relationships for SC and its constituent elements. Finally, and not surprisingly, all of this would be a pipe dream if not properly
resourced, and the Military Departments (ie Department of the Army, etc)_and
Combatant Commands (ie. Central Command or CENTCOM) must be provided
the means to organize, train and equip capabilities for this.[25]

    While the SC Roadmap provided the DoD with authoritative guidance with
which to shape capabilities and operations; the interagency organizations had no
such guidance, yet there is hope, that it may have an even broader plan. In October
2006, the Under Secretary of State for Public Diplomacy and Public Affairs, Karen
Hughes, circulated for coordination a memo entitled U.S. National Strategy for
Public Diplomacy and Strategic Communication, under her hand as chair of the
PCC for PD/SC. This was a much longer and more strategic document that set
forth three strategic imperatives that would guide American PD and SC programs.
The first of these was the importance of presenting a positive vision of hope and
opportunity, rooted in basic American values. Next was the need to isolate and
undermine violent extremists, while the final imperative was to nurture common
interests and values and emphasize that they cross cultures, borders, and creeds.
The draft strategy went on to identify critical influencers who are able to reach
"strategic audiences" and "vulnerable populations". It also emphasized the need
for interagency coordination, because every arm of the US Government has an
urgent mission in this arena. Its "action plan" was based on the three strategic
imperatives, and nearly 40% of the entire document was devoted to specific and
detailed plans and proposals. Finally, the draft strategy examined several critical
elements of communication, such as broadcasting or public opinion analysis, that

would be necessary supports for a successful strategy, and it emphasized the need to be accountable for operations and to gauge whether any specific plan or program was being successful.[26]

This plan was broad and inclusive, a major step forward that went well beyond anything that had existed previously. One major improvement over earlier efforts was that the PCC charged with developing this strategy was not "co-chaired" and thus did not suffer from divided leadership, and was instead led by one person, indeed led by one of the most influential members of the Bush Administration. Karen Hughes' unique power stemmed from her relationship with the President and her position as one of his key advisors, so that her guidance always had an "ex cathedra" aspect to it. This provided a unique "window of opportunity" in which perhaps real progress could be made before the pressures of the pending 2008 elections and an administration changeover in 2009—regardless of which party was victorious—brought efforts back from full speed.

There were weaknesses in the plan, however. One was its insistent focus on the Moslem/Islamic world. While that was quite normal in one regard, especially in its connection to the "Global War on Terror", in other ways that emphasis was unfortunate, because there were other areas of the world, Latin America, Asia, sub-Saharan Africa, to name just three, in which America needs to be fully engaged in support of vital national interests. Another area in which the plan was even more inadequate was the almost perfunctory paragraph on resources. Instead of a powerful and compelling call for greatly increased resources with which to wage the "war of ideas", and a detailed explanation of how those resources would enable the United States to advance its interests, the strategy provided merely a weak one-liner about the need for "increased support". In a fiscal environment in which every dollar has several worthwhile programs calling for it, such a weak request has virtually no chance of actually gaining the needed resources, which will unfortunately probably be the quick demise for this noble effort.

Organizational Changes

If one thinks that there has only been major additions in IO to policy, then they would be mistaken. Organizationally the landscape of IO has shifted dramatically as well. One analogy often used to describe the changing role of IO from an organizational perspective has been suburbanization of the warfare area. Ten years ago, with the huge emphasis on the revolution in military affairs, and the introduction of IW, grand themes and terrible scenarios were described in great detail. These included the like of "Electronic Pearl Harbor," "CyberWar," and other similar threats that provided a degree of "hyped emphasis", but while helping to introduce the vulnerabilities associated with IO, these "chicken little type hysteria" also had the unfortunate effect of desensitizing personnel to the real dangers that often tend to be mundane and technologically complex. For example, early descriptions of cyber attacks often foretold of massive panic as hackers brought down the power grids in the United States. However when this actually happened on the 14th of August 2003 in the northeast portion of the United States due to a fault in a power plant, it wasn't panic that ensued, but instead millions of people were relieved that it was in fact only a technical hitch and not a terrorist attack, and their bemused attitude, as they migrated their way home in a hot and power-less day was a perhaps refreshing demonstration of peoples' resilience. So in fact, the Electronic Pearl Harbor *did* occur as predicted, not due to a cyber attack, but instead due to a mechanical error. It is this movement from the Wild West attitude to a more operational or "surburbanized" effort that probably best reflects the overall theme of this section and book. No longer can agencies develop IO solutions alone or in a vacuum, and so what will become increasingly apparent to the reader is that changes to policy and organizations tend to become less profound but more detailed and with depth and substance as time passes.

IO: Organizational Change

As IO has become more accepted and operationalized across the DoD, resources have been provided and the offices mentioned in the previous text have been broad-ened and enlarged to accept these additional responsibilities. Yet all changes have not been exactly straight-forward. A few years ago, IO within the Joint Staff was the responsibility of the J-39, one of the sub-offices of the J-3 Director of Opera-tions. By 2002 the J-39 had evolved into the J-3's Deputy Director for Informa-tion Operations (DDIO). By 2006, however, that same individual (a brigadier general) had assumed the mantle of Deputy Director for Global Operations (DDGO), and oversight of IO became only one of several responsibilities, so that the direct oversight of IO subsequently devolved downwards within the DDGO. This is in contrast to the growing scale and scope of IO in the operational world, in which the planning and conduct of IO features ever more importantly in US military operations.

Within the OSD structure, the organization of IO has also been fragmented. For the last few years, IO was the responsibility of the Assistant Secretary of Defense for Command, Control, Communications and Intelligence (ASD/C3I). By 2006, however, at least four major OSD offices "owned" significant segments of IO. The old ASD/C3I has become ASD/NII (for "Networks and Information Integration") and is now responsible for only the IA and CND aspects of IO. Its intelligence responsibilities have been transferred to a new office, the Under Sec-retary of Defense for Intelligence, USD/I, which has the overall IO portfolio as well as specific responsibility for Military Deception, Operations Security (OPSEC), and CNO (specifically the attack and exploitation elements). The As-sistant Secretary of Defense for Acquisition, Technology and Logistics (ASD/ATL) has the EW mission, while the Under Secretary of Defense for Policy, USD/P, has oversight of PSYOP. While some may see these organizational develop-ments as the "balkanization" of IO, others see it as IO's "normalization", and argue that IO could never realize its full potential and impact as long as it was stove-piped within a single section of OSD, and that these organizational devel-opments actually work to embed this capability more fully across the entire DoD structure.

What has been perhaps the most significant and visible change from an organizational aspect is in the placement of IO in the operational military tasks assigned to the Combatant Commanders. While previously IO was the responsibility of Space Command in Colorado Springs, in 2003 a major change in the Unified Command Plan (UCP) changed this and assigned overall support of IO to Strategic Command (STRATCOM) at Omaha, Nebraska. This was done on the belief that the latter organization was better equipped to conduct IO on a world-wide basis, utilizing the additional support that this agency had received after 9/11. When General Cartright took command of STRATCOM in 2005, he set in motion a command-wide reorganization in which IO was a central issue. General Cartwright created and delegated operational responsibility to a set of subordinate commands called Joint Force Component Commands, or JFCCs. To better link STRATCOM's capabilities to those of other important organizations, General Cartwright created the JFCC for Network Warfare and placed it at Fort Meade in Maryland. Likewise, the operational Joint Task Force for Computer Network Operations (JTF-CNO), which evolved from disparate offensive and defensive components was not only merged together, but also became the JFCC for Global Network Operations. Located in Washington, D.C. at the Defense Information Systems Agency (DISA), the JFCC-GNO is now the preeminent IA center for all of DoD. Likewise the new JFCC for Global Strike was also stood up and this includes Joint Information Operations Command in San Antonio, Texas.

From a service perspective, a number of changes have also occurred, in addition to those mentioned previously. The U.S. Army has renamed the old Land Information Warfare Activity (LIWA) as the 1st Information Operations Command (1st IOC), and has increased its staff dramatically since the terrorist attacks. Their IA center has changed as well and is now called NETCOM. Coordinating with both the 1st IOC and JFCC-GNO, NETCOM is basically the Army's Computer Emergency Response Team (CERT). This approach is similar but slightly different than the Navy. The Navy's Fleet Information Warfare Center (FIWC) became the Navy Information Operations Command (NIOC) within the Naval Network Warfare Command (NetWarCom), as these commands have evolved to conduct their day-to-day missions. The Navy Computer Incident Response Team (NAVCIRT) was spun off of FIWC in 2004, but with this latest reorganization in 2006, both of these units are now firmly a part of the larger NetWarCom agency. Different from both of these services is the Air Force, which still retains the Air Force Computer Emergency Response Team (AFCERT) as part of the Air Force Information Operations Center (AFIOC). The AFIOC itself has benefited from the Air Force's growing emphasis on IO, with the creation of the 67th Network Warfare Wing at San Antonio, as well as the 318th Information Operations Group, with its six component squadrons, provides operations capabilities ranging from testing technology to the development of tactics, techniques and procedures. Without question, however, the most significant organizational development within the Air Force has been its move "into" cyberspace and its decision to nominate its storied 8th Air Force to become its "Operational Cyberspace Command", to eventually stand alongside its

Air Combat Command and Space Command at the major command level.[27] Taken together, all of these changes show a significant effort to operationalize and normalize IO within the traditional structure of the military services.

The events of 11 September were also a tremendous wake-up call for the Bush administration and how it conducted IO at the executive level. After these attacks, the State Department was looking to the executive branch and the NSC for guidance on building an organization to support a strategic information campaign. Unfortunately however, leadership was slow in forming. In the immediate aftermath of the terrorists' strikes, there was a significant amount of confusion within the government. For the first five to six weeks at the NSC, there was an absence of knowledgeable, experienced people to deal with strategic influence campaigns, as well as the normal intra-organizational discontent and turf battles. At that time, the Clinton-era NSC document, PDD-68 International Public Information (IPI) had been effectively muted, so there was no office dedicated at NSC to conduct a strategic perception management effort. The Joint Staff ended up during major portions of this crucial period simply contracting out their perception management campaign to the Rendon Group, a civilian company that specializes in strategic communications, under a contract with ASD/C3I.[28] Gradually, as the campaign on terrorism continued throughout the fall of 2001, a number of influence plans and strategies were developed to create a working operational group, yet the hoped-for NSPD still remained in a holding pattern within the interagency process. In November 2001, in accordance with NSPD-8, which established the Office of Combating Terrorism and outlined General Wayne Downing's roles as Deputy Assistant to the President and National Director and Deputy National Security Advisor for Combating Terrorism, a new position of Senior Director for Strategic Communications and Information was filled by a very experienced Army PSYOP, retired Army Colonel Jeffrey Jones, former commander of the Army's 4th Psychological Operations Group (4th POG).[29]

Likewise during the immediate aftermath of the terrorist attacks, Alstair Campbell, the Communications Director for British Prime Minister Tony Blair had suggested to Karen Hughes, Communications Director for the Bush administration, forming a series of Coalition Information Centers (CIC) to concentrate on getting the pro-American message to the world media. Eventually three were set up, in Washington, D.C., London, and Islamabad. The facility in Pakistan was actually an old USIA building. These groups perform admirably, focusing on public affairs and public diplomacy, yet some critics argue that these organizations concentrated on U.S. domestic partisan politics instead of focusing on the set of global audiences now accessible via a 24-7 news environment. Other critics have argued however, that these CICs generally worked well informing domestic and foreign press within their time cycles during the early phases of OEF, and they also eventually received a U.S. government spokesman who could speak Arabic and thus appear live on the Al Jazeera TV station. Of course looking back, one cannot be sure that this really was a success story. One must ask the question of why it took so long for Ambassador Christopher Ross to appear on Al Jazeera.

This may have been because the White House was slow to see the need for United States' presence on Al Jazeera until external pressure became so bad that it actually forced Colin Powell and Condoleezza Rice to appear on Al Zazeera using translators. In fact, Al Jazeera constantly invited them for interviews early on, but these invitations were rebuffed and Al Jazeera was blacklisted from early White House press conferences. Eventually the response was changed, but it should have been recognized much earlier.[30] Foreign media also needed to be addressed, and the fact that it took so long to make key U.S. government personnel available was rather depressing and was perhaps an indication that at the highest levels the US Government did not understand the true nature of this new battlespace.

As the war effort in Afghanistan continued, much of this effort was minimized with the shutting down of the Islamabad facility. Before she left the Bush administration in its first term, Karen Hughes formed the Office of Global Communications (OGC), ostensibly to force the public diplomacy community resident within the DoS and in the field to do a better job of explaining overall U.S. policies. Created out of frustration with the perceived lack of effort at Foggy Bottom this office coordinated with the interagency Global Communication Strategy Council. An evolutionary process and a follow up to the CIC, this White House office coordinated with the NSC, in a quid pro quo relationship. Yet the departure of Hughes and later General Downing from the Bush administration probably spelled the ultimate demise of the OGC and further White House Strategic Communication efforts, in the early post-9/11 timeframe.[31] There are those, however, who don't believe this was OGC's mission at all, and believe its real mission was to be the influence arm of the White House and get the President's message out as an element of his reelection campaign. While this would be a normal and understandable objective of any White House-based communications effort, suspicions remain that the then-director of White House communications, Karen Hughes, quickly acted in early 2002 to put the new SC PCC on hold because of fears that it would interfere with this mission. The fact that shortly after the election of November 2004 this Office of Global Communication quickly and quietly ceased operations supports this interpretation.

The emphasis on the domestic audience can have negative effects in other ways, too. There is a lack of understanding about what words or phrases mean to other audiences: some may be instantly hostile to an Islamic audience, while others may have an impact poorly understood by Westerners. Axis of Evil, Infinite Justice, and Crusade are great examples of Bush administration's public diplomacy missteps. The White House did not collaborate well with State Department specialists who understand the implications of such phrases; these actions and words have seriously hurt the Bush administration in its global war on terrorism. Some quip that a serious review of Samuel Huntington's *Clash of Civilization's* is not out of the question. On the other hand, the use of commonly used Islamic terms to label our adversaries may have a negative and unintended consequence. Including suicide bombers and terrorists under the label "jihadists" may actually be seen as legitimizing them and their actions. We use labels and terms in many cases because they are easy and in the common lexicon, yet we don't understand how they appear and what they mean in other cultural contexts. In a "war of ideas", words can be ammunition.[32]

The IO organizational changes at the interagency level only got more convoluted as the Global War on Terrorism continued.[33] The J-3 Director of Operations on the Joint Staff formed the Information Operations Task Force, led by the J-39, to be responsible for IO, but that group was more technically oriented, so there was still a role for the DoS in the diplomatic arena.[34] A Strategic Information Core Group was also formed within the interagency structure, but overall, the general consensus was that not much was accomplished with this organization because they were never empowered or recognized by the major departments to possess the ability to get things done. Into this atmosphere of OEF and the ongoing war in Afghanistan, in November 2001 the Office of Strategic Influence (OSI) was established by the DoD in an effort to coordinate its strategic perception management campaign and because of a perceived leadership void, with the Assistant Secretary of Defense for Special Operations/Low Intensity Conflct (ASD/SOLIC) in the lead. It appeared to be placed to work well, because one it had financial resources, and two it was also a DoD organization, yet it quickly ran afoul of two critical interagency IO organizations.[35] For the OSI group had been placed at DoD, not at State Department's Bureau for International Information Programs (IIP) because some believed that its more operational tasks may have been more easily accomplished from within the DOD. This ultimate rejection of PDD 68 may have stemmed from the overall belief that the strategic perception management campaign had been wrongly placed by the Clinton administration, and that instead, an office should have gone to the DoD or NSC instead. The OSI organization was comprised mostly of personnel with PSYOP and civil affairs backgrounds, with a mission to respond to and negate hostile propaganda, using mostly human factors and a little technology.[36]

At a meeting on 16 February 2002, Secretary of Defense Donald Rumsfeld approved the office, however the senior DoD Public Affairs official Victoria Clarke did not concur, and her opposition manifested itself almost instantly. On 19 February the first article critical of the new organization appeared in the New Yorks Times while both Rumsfeld and Clarke were in Salt Lake City, Utah at the Winter Olympics. It was reported that Rumsfeld was livid but could not do much due the political concerns created by the allegations that OSI would lie to the media to conduct disinformation campaigns.. As satirically reported by Mark Rodriguez in the *Washington Post* electronic journal *Insight*, the demise of this DoD office was a political turf-battle with Clarke leading her own disinformation campaign to retain control of all PA efforts, exactly the charge she made to the press about OSI, which was later investigated and proven unfounded.[37] Politically embarrassing to Secretary of Defense and the President, it was very comical to watch the government officials deny the need for an office in the United States to conduct strategic perception management campaigns. Every nation participates in these activities, but almost all deny their existence. Even foreign news agencies put a satirical touch on their reporting as they watched the American officials attempt to explain away the obvious.[38]

All of these organizational shifts with regard to SC allude to a question that has arisen over the last five years, namely where should a strategic perception management campaign office be located? PDD-68 put the International Public Information (IPI) activities at the State Department in 1998 where it foundered for two years due to lack of budgetary authority, manning, and empowerment. In addition, the IPI group was also hampered by the interagency process. While the draft NSPD has repeatedly recommended the need to embed the strategic perception management capability in an office in the NSC, the DSB-MIB in 2001 reiterated the desire to keep the authority at DoS. The argument for keeping the Policy Coordinating Committee (PCC) at NSC was basically because that organization is in a steady state. The NSC is by definition, the single organization within the U.S. government responsible for turning interagency positions into recommendations to the President. It looks at international affairs and foreign audiences in an operational manner, which was greatly missing from the IPI way of doing business. So there is strong logic behind this argument as well. The counter-prevailing suggestion for putting the PCC in DoS was led by David Abshire, who believed that a Tom Ridge-like figure was needed to drive the program.[39] However, there is also a concern that any SC effort led by the DoS will be focused more at PD/PA than strategic influence issues.

Yet all of this effort was eventually overcome by events. With the initial departure of Karen Hughes from the White House in 2002, most of these activities lost their momentum. For it was, after all, Karen Hughes who made the CICs happen during the early stages of OEF. She understood how effective public diplomacy could be on the war on terrorism. The CICs were so successful during the fall of 2001, mainly because of the President's influence, and also because there were effectively no constraints. In effect, they didn't have to filter information through a number of layers of bureaucracy. Normally, Congress is very concerned with the Smith-Mundt Act, an early Cold War-era piece of legislation that prohibited the delivery to the domestic American populace of any foreign-targeted information.[40] As the events surrounding the Office of Strategic Influence debacle of early 2002 indicated, the widespread concern towards activities of the State and Defense Departments may have not been the case when it comes to the White House. With the creation of the OGC and its assigned mission of explaining the United States policies the White House felt a great need during OEF to expand their frame of reference, for example to influence those Islamic nations and populations that reject out of hand any information coming from western sources. This theme was emphasized by David Hoffman of Internews Network in his *Foreign Affairs* article, "Beyond Public Diplomacy."[41] In asking the quintessential question "How can a man in a cave out-communicate the world's leading communications society?" he thus strikes a cord for more concerted strategic communication efforts by the U.S. government. Therefore the DoS still needs to enlist moderate Arabic nations to help in this project, but this desire runs into the roadblock of how current American efforts in Israel/Palestine conflict are seen across the Islamic world—and exploited by Islamic radicals, sometimes via overt disinformation—as clear evidence of a "US-Zionist alliance". The conflict in southern Lebanon in summer 2006 merely added fuel to this fire. Often the U.S. government does not necessarily see the connection, but the entire Arabic world instantly does. So now the White House is even losing out on trying to get the moderates to push our message. Plus the debacle concerning the OSI in February 2002 also stalled many of the subsequent Bush administration's attempts to develop a strategic communication effort, and essentially this controversy put the NSC's Strategic Communication PCC on hold, until the creation in April 2006 of the new Public Diplomacy/Strategic Communication PCC, chaired by Karen Hughes as discussed previously.[42]

Thus the mission and structure of the new PCC constitutes an attempt by the Bush administration to develop a long-term capability to conduct public diplomacy and strategic communication. While there is still no overarching U.S. government strategy for Strategic Communication, despite the fact that the White House has had a Counterterrorism Information Strategy since December 2001, there can be little doubt that the proposed strategy circulated for coordination by Karen Hughes in late 2006 will become the long-sought government-wide effort. The irony is that it was less than five years ago that the USIA was dismantled, and its functions shifted under the greater umbrella of the DoS. In fact, as mentioned previously, representative Henry Hyde (R-NY) proposed numerous times the reconstitution of that agency, in his legislation to bring back capabilities that had so recently been diminished, for much of this legislative proposal mirrors efforts by the DSB-MIB working group. While the State Department did not agree with this concept, in the long run the new structure suggested by the Karen Hughes-chaired PCC may go even beyond what existed previously in terms of a strong centrally-led influence and communication program.[43] Therefore, the demise of the USIA may have contributed more to the failing of PDD 68, and thus the need for a new structure and capability to conduct global influence than any other action to date.[44]

For in the end, it is not a new organization that will drive a strategic communications effort but instead a shift in the mindset of the White House and the NSC. The need to push senior officials to conduct briefings at 0700 EST, to match Middle Eastern news cycles, or to ensure U.S. Arabic speakers are available on Al Jazeera, are becoming much more accepted and understood methods of doing business. These ideas are now conventional wisdom as the value of strategic communications rise within the Bush administration. To be effective, one cannot just think in news cycles (24/7 around the world), but instead also in decades, for example, expanding exchange programs such as the Fulbright Scholarship program, so that the U.S. government can be much more effective in a strategic management campaign. This would be an example of one of Karen Hughes' "Four Es" of Public Diplomacy: engage, exchange, educate and empower. In effect, there needs to be an issues agenda versus a value agenda. We need to take a short- and long-term approach to these problems, but it must also be led from the top down, with full

White House and NSC leadership to ensure full interagency participation.[45] The Administration has repeatedly tried "talked the talk" of PD and SC, and at all levels, from Vice President Cheney through Secretary of State Rice to Under Secretary Hughes, quotes and sound bites referring to the need to do these tasks better abound. But, what is really needed now is "walk to match the talk", ie real evidence of resources, organizations, people and operations that enable an effective long-term SC campaign. It is only then that a true strategic perception management campaign will succeed, and the power of IO be realized by the United States.

Conclusion

What all these policy developments and organizational changes have recommended and attempted to explain is a much greater emphasis on the use of the information environment across the spectrum of national security activities, from perception management capabilities by the federal bureaucracy to engage in strategic operations in the Global War on Terror to securing critical information infrastructures against terrorist attack to military employment of the full range of IO's core competencies. The publication in late 2003 of the Secretary of Defense's the IO RoadMap was a major step forward in the development of this warfare area within the DoD. The result of five years' cumulative efforts of updates and changes based upon real-world operations and missions conducted by the military around the world, the IO Roadmap concentrated more on the traditional aspects of IO including and in many regards was seen as a revalidation of the old concept of C2W. Subjects such as PM, SC, PD and influence campaigns were minimized in the Roadmap. While in one view, this new policy could be considered a failure because its narrow focus on traditional areas of IO, it once again highlights the huge mismatch between the strategic transformational promise of IO doctrine and the operational reality of how the Defense Department tactically conducts information activities and campaigns, for in reality the IO RoadMap may very well have been the best pragmatic solution. The new Joint Doctrine for IO, Joint Pub 3-13, built on the IO Roadmap to make another major step forward, and it marked the growing comfort level with the embedded role of IO within basic military strategy and operations. The year 2006 may come to be seen as the period when every aspect of IO in the national power structure moved forward. The IA world saw the publication of the *National Infrastructure Protection Plan*, the SC world saw the development of a log-awaited draft strategy, and the military now has approved joint doctrine on which to base plans and operations. The real question of course is whether this growing set of policy and guidance documents and proliferation of IO related organizations indicate a greater understanding by the U.S. government as a whole and its constituent elements about the power and capability of information as an element of power in this new era. This we will have to wait to see.

---

* This chapter reflects the authors' opinions and should not be taken as official U.S., DoD, or NDU policy.

CHAPTER 3

Perception Management: IO's Stepchild

Pascale Combelles Siegel
President, Insight Through Analysis

The information revolution has changed the American way of war. Conflicts have increasingly become struggles over information and information sys tems. Effective collection and collation of information provide a means to operate inside enemy decision cycles. Physical destruction increasingly targets communication and transmission nodes and means. Effective intelligence is ever more critical for tasks such as striking time-critical targets—as is often said, "precision weapons need precision intelligence."[1] Joint Publication 3-13, "Information Operations" (IO), published in October 1998, examined some of these changes and sought to provide order to the military's approaches to IO. Joint Pub 3-13 defines IO as "actions taken to affect adversary information and information systems."[2] The ultimate offensive goal of IO is to target information and information systems in order to corrupt the adversary's decision-making process and impair his ability to act in his own interest. While IO is just one tool in a commander's arsenal, it is a tool receiving increasing focus as per publication of Joint Pub 3-13.

IO encompasses a large spectrum of specialties, from computer network attack (CNA) and defense (CND), to deception, to electronic warfare (EW), to psychological operations (PSYOP) and, as support elements, civil and public affairs (CA and PA). In general, the Department of Defense (DoD) has focused most of its attention on technology and intelligence. It has committed most of its resources (both financial and human) to improving intelligence, securing networks, and digitizing the battlefield. Those dominating IO tend to come from technological, intelligence, and communications backgrounds. Within this context, a critical arena often ends up as the ugly stepchild in the equation: perception management (PM). This article seeks to bring a different light to the PM's importance in facing twenty-first century security challenges.

After a brief definitional discussion, this article turns to perception management's importance and ways in which the United States (and western world) are vulnerable to adversaries' PM operations. An analysis of why PM efforts, although beefed up in the past few years, have fallen short is included. It concludes with paths to improve PM capabilities.

Definition

Even if inadvertently and—as will be discussed—with haphazard coordination, perception management is widely practiced and has become an increasingly important tool of warfare and peacemaking. The Department of Defense defines perception management as follows:

> Actions to convey and/or deny selected information and indicators to foreign audiences to influence their emotions, motives, and objective reasoning as well as to intelligence systems and leaders at all levels to influence official estimates, ultimately resulting in foreign behaviors and official actions favorable to the originator's objectives. In various ways, perception management combines truth projection, operations security, cover and deception, and psychological operations.[3]

While quite useful, this definition focuses on deliberate actions to influence rather than the totality of one's activities and others' perceptions of you in terms of influencing their views. Thus a more robust, inclusive definition might be worth exploring.

This author defines PM as the ability to shape worldwide perceptions in one's favor to foster compliance and facilitate mission accomplishment. A critical part of PM is the effort to understand others' perceptions and basis for those perceptions as a path toward understanding how one might then influence them.[4] The basis for perceptions includes many issues that are not just outside DoD, but U.S. government (USG) control—such as television and cable sitcom or even the millions of personal home pages of American teenagers that are accessible to anyone with Internet access. In other words, a critical part of PM is to need to map and analyze the perception influencers on a constant basis. With the beginnings of such an understanding, one can then turn toward the orchestration of a delicate minuet of activities designed to manage views.

Within the DoD, this minuet must deal with the coordination and synchronization of various command specialties, including public affairs, intelligence, psychological operations, civil affairs, and some Department of State (DoS) specialties such as public diplomacy, international public information, and international broadcast systems.[5] Specifically, PM seeks to:

- build and preserve public opinion support (at home and abroad) to gain and maintain legitimacy,
- communicate intent and objectives to hostile and/or third parties to establish a high degree of credibility so they fully understand the consequences of their actions,

■ influence the attitudes and behaviors of the local populations so they act in accordance with U.S. objectives.

PM must target many audiences. Domestic audiences require information about an operation's rationale, risks, and benefits, as without public support, democracies cannot sustain military engagement. Meanwhile, adversaries' and third-party perceptions have to be managed so that they reorder their priorities and strategies in accordance with U.S. goals and objectives. Achieving both goals in outright warfare (or Operations Other Than War, OOTW) is fostered by active perception management efforts.

Perception management is crucial across the conflict spectrum from peacetime to high-intensity conventional warfare. In peacetime, PM is critical for two of the four Department of Defense objectives defined in the 2001 Quadrennial Defense Review—to dissuade potential adversaries and deter them from acting against U.S. interests. PM is a critically important dimension of crisis resolution as well. What, after all, is gunboat diplomacy but PM using military forces? Many peace operations commanders have emphasized the role of information as a nonlethal weapons system. Finally, PM is a critical tool in combat missions. Propaganda efforts during World War II were critical to lowering the morale of Japanese forces. During Desert Storm, Central Command's (CENTCOM) press briefings played a key role in conveying messages to Saddam Hussein and his high commanders while PSYOP operations led more than eighty thousand soldiers to surrender without shooting at U.S. forces.[6]

Why Perception Management Matters

Military operations are, in fact, influence operations designed to persuade parties into a preferred course of action through the threat and/or use of force. At its core, the use of force aims to convince adversaries or third parties to act in accordance with U.S. interests and goals. For example, while the nuclear weapons dropped on Hiroshima and Nagasaki destroyed much Japanese infrastructure, the intent was to induce the Japanese to surrender—which they did. Infantry patrols in Kabul are prepared to engage in combat but are intended, principally, to maintain order by assuring residents of stability and deterring potentially violent elements from engaging in violence. From the tactical to overall strategic environments, PM lies at the core of military activities even if not command processes.

In this environment, credibility and legitimacy are key. If the domestic and international public opinions are not convinced that the operation is fully justified, their support will slowly erode, maybe to the point of undermining the sustainability of U.S. commitment. In most twenty-first-century warfare environments, legitimacy will perhaps be *the* center of gravity for U.S. warfare. Within domestic and non-adversary audiences, legitimacy is the perception of an appropriate legal, moral, and ethical basis combined, often, with a belief in the necessity of military action. Legitimacy stems from the mandate, the respect for the law of war, and regard for humanitarian principles. Legitimacy is key because it sustains popular support and will to fight without which democracies cannot sustain military action.

While legitimacy is a crucial underpinning to maintaining an operation, credibility is key within both friendly and hostile audiences. Adversaries must believe that the United States has *both* the *capability* and *will* to act—their perceptions must be managed.[7] Saddam Hussein fundamentally questioned whether the United States had the capacity to intervene in the Middle East prior to the invasion of Kuwait due to deployment issues and questions within the United States about U.S. capabilities to face Iran—an enemy he had defeated. With the threat of the "Mother of all Battles," he showed a belief that the United States lacked the will to fight and endure massive casualties. Similarly, prior to 9/11, it seems clear that Osama Bin Laden and his Taliban protectors misunderstood U.S. capabilities and will to act against them in the face of a massive terrorist attack.[8]

Why Is There a Vulnerability?

Targeting perceptions have always been part of warfare—from screaming tribal warriors pounding spears against shields to "Tokyo Rose" cooing to American servicemen across the Pacific. Recent conflicts, from Vietnam to Somalia, and Kosovo to Afghanistan have taught adversaries that perception management represents not just a useful but also potentially the most effective weapon against U.S. military operations. Developments in modern technologies have magnified the reach and power of hostile perception management operations and increased American vulnerabilities.

A GLOBALIZED MEDIA

Media operate on a worldwide basis. Although most U.S. media have significantly decreased their worldwide presence for financial reasons, stringers, technical reach (see Internet below), and commercial agreements with local outlets have created a virtual presence. (CNN and Al Jazeera reached such an agreement years ago, whereby both networks have rights to access and use each other's video feeds. After 9/11, Al Jazeera was the only broadcast media presence in Kabul which meant that CNN had a lock on footage from Kabul until pressure from other media outlets led to an open sharing of Al Jazeera footage in the U.S. market.)

Reach has changed. Broadcasts are no longer local or confined to traditionally defined domestic markets. Satellite, cable, and Internet provide worldwide broadcast reach for any media outlet. Media globalization ensures that events taking place virtually anywhere can have worldwide impact. By virtue of technology and commercial agreements, local news can become international news and suddenly impose its urgency to all worldwide leaders. Conceiving of separate messages for different audiences can be dangerous as worldwide reach means that anyone might see any message. This phenomenon affects everybody. For example, Saudi Arabia was not intending for the American public to hear of government-run television programs running vitriolic anti-Jewish commentaries. However, the West is all the more vulnerable because of its greater openness and higher journalistic standards.

NEVER-ENDING NEWS CYCLE AND EVER-PRESENT NEWS REPORTING

Technological changes have helped redefine the news cycle. Between cable and the Internet, all-news television and radio stations, the news is constantly on and there is tremendous information space to be filled. As nature abhors a vacuum, the expansion creates tremendous opportunities for those who want to speak out to be heard. In the search for scoops and market advantage, standards for verifying "facts" all too often fall to the wayside. The "vacuum" also fosters an even greater expansion of punditry, which is typically at its worst when hard facts are the least available. With little true news to report, media outlets turn out to "experts" to fill in open time. Sadly, the marketplace proliferation creates opportunities to give credibility to ever less credible "experts." And, as marketers understand, people give great credence to the words "as seen on TV."

REAL-TIME INFORMATION

With the exponential growth of media outlets, all-information networks, round-the-clock operations, and the Internet, the news cycle has shrunk to a constant stream of information. Thirty years ago, officials dealt with media deadlines. Newspapers were going to print once a day (either in the early afternoon or in the late evening), radio had two major news programs a day, and television networks had the six o'clock evening news. Public affairs organized around their deadlines.

The twenty-first-century news business is real-time, rather than deadline, dominated. The pressures of competition and the need to fill an ever-expanding information space means that increasingly being first matters more than being right. As seen on TV, reporting live too often seems more important than actual content. In that context, rumors, half-truths, and unchecked information can quickly become news. This can also place intense pressure on officials to react to stories before they know all the facts. Due deliberation is not fostered by this environment. Worldwide reporting of the Israeli "massacre" of Palestinians at Jenin led many to issue statements and, in too many cases, act in reaction. There was no massacre—it was effective Palestinian and ineffective Israeli perception management at work.

THE INTERNET

The Internet compounds these effects. It has created a space where information can be easily acted on and disseminated. On the Internet, information takes a life of its own. False information lingers and lives on. Alongside the proliferation of news outlets via satellite and cable, the Internet is creating an environment where it is increasingly difficult to assume that people have common *weltanschauung*—individuals might spend more time with people and information sources from halfway around the world than with their neighbors or domestic press.

The Internet also adds tremendous pressure on traditional reporting. The Internet may simply bypass restraints decided by one media organization. For example, as U.S. troops arrived in Uzbekistan, CNN adopted a no-say policy, meaning it did not report the event. However, a Pakistani reporter posted the information on the Internet.

EVER-EASIER NEWS CREATION

The creation of news reporting has never been cheaper and easier. Under the right set of circumstances, a low-quality but newsworthy video can have a surprising effect. The October 1993 video of an angry Somali crowd dragging the body of an American serviceman in the streets of Mogadishu, perhaps more than the battle itself and the U.S. casualties, put Somalia on top of the political agenda in Washington. Similarly, it took just a few VHS tapes—the same type of cheap technology that fostered Ayatollah Khomeini's rise to power over two decades ago—thrown into the car of an Al Jazeera journalist for Osama Bin Laden to reach a worldwide audiences and energize the Arab streets against the Global War on Terror.

These developments create tremendous challenges for perception management efforts. Adversaries have near instantaneous access to audiences at incredibly low cost. The structure of the media environment and democratic institutions can make it easier for adversaries to access and influence worldwide reporting. Not everything is bleak, however; these developments also create tremendous opportunities to turn the situation to our advantage. And, an imperative exists to take advantage of these opportunities.

Perhaps the most compelling argument for paying attention to perception management is that America's adversaries, who do not really have the means to defeat the United States conventionally, have used and will continue to use perception management to weaken American resolve and commitment. In other words, if we don't do it, others will do it for us.

PM is indisputably a serious asymmetric threat to military forces from democratic nations, notably U.S. forces. No nation has the capability to defeat the U.S. military in conventional combat. Desert Storm demonstrated to potential U.S. adversaries that the U.S. military's strength, resources, equipment, and professionalism make it essentially immune from defeat in a traditional sense. Arguably, U.S. defeats from Vietnam to Lebanon have had something to do with "losing the hearts and minds." As suggested by the brief case studies below, even unsophisticated, ill-equipped, and ill-trained adversaries can successfully use PM to erode the West's will to fight or to make peace.

Insufficient Efforts to Date

The U.S. military has long paid attention to the importance of domestic public support for sustained military engagements. In 1983, the Weinberger doctrine established the need for domestic public support as one of six prerequisites before going to war. Meanwhile, since Desert Strom, PSYOP has regained its reputation, shattered as a result of Vietnam, as a useful tool of warfare. Since then, the military has undertaken major steps, both in terms of doctrine and professional training, to address both the importance and contribution of perception management to an overall military effort. As years and experiences passed, the military has reviewed and improved its PA and PSYOP doctrines.[9] It has studied how civil affairs operations and force posture impact the adversaries' and third parties' perceptions and shape their behaviors. It has sought to apply effect-based-operations principles to both PSYOP and public affairs.

As outlined in the introduction, DoD took a major step in the 1990s with the emergence and refinement of the information operations field and related doctrine. Information operations use information to corrupt an adversary's decision-making process to the point to impairing his ability to fight and make decisions in his own interest, while protecting one's own cycle. As a result of these efforts, the U.S. military now has a doctrine and established procedures to better implement the promises of information operations. The DoD has also significantly increased selected capabilities and reorganized resources to achieve a better synergy.

For both cultural and philosophical reasons, perception management tools have not fared as well as their technical counterparts. PA and PSYOP forces have not increased substantially, despite increased operational needs. Many organizations own a part of perception management and this renders cooperation efforts more difficult. Getting the right equipment to right people at the right time remains a challenge, especially for PSYOP. Money is always tight for operators to bring new initiatives to life. The relationship with other USG programs is always an issue and in many cases, the USG has not been able to make a viable case to the world. The following paragraphs provide some case studies that point to some U.S. difficulties in this arena, past and present.

THE SOMALIA CAMPAIGN (1992–1993)

During the UN operation in Somalia, perception management was a major battleground. Before the UN mission deployed in May 1993, Aideed's faction used Radio Mogadishu to delay, shape, and undermine the UN operation. Radio Mogadishu repeatedly accused Italian businessmen of dumping toxic waste in Somalia. Subsequent reports accused the Russians of using UN-marked planes to deliver weapons and ammunitions to Al Mahdi (Aideed's rival). The accusation was false, but it worked to weaken Russia's ability to become a major player in the UN operation and served to discredit Al Mahdi.

With the UN deployment, Aideed sought to gain U.S. support *and* undermine the UN reconciliation process. To that effect, Radio Mogadishu consistently played anti-UN rhetoric while praising U.S. actions. (This played to the U.S. domestic audience, including many in Congress, who were deeply distrustful of the UN). Aideed's aides fed information and sound bites to American journalists. These included the oft-used comparison of Aideed with George Washington as generals who were fathers of their countries. Aideed's faction also staged anti-UN rallies for the benefit of the international media. This wasn't hard as most correspondents stayed at the K-4 hotel in Aideed territory. Was it a coincidence that most anti-UN rallies were within shouting (or, at least, shooting) distance of the K-4 circle?

As an SAIC analysis of Somalia concluded:

The international media is the most powerful offensive and defensive information warfare asset available to lesser-developed countries and/or parties to a conflict . . . . From the perspective of a less technological advanced "opponent," the media provides real-time information about U.S. political deliberations and military preparations and provides a high-tech capability that can be used to disseminate verbal and symbolic propaganda against the U.S. population.[10]

As the fighting escalated and the international community decided to remove Aideed, his disinformation campaign became increasingly hostile to the United States. Radio Mogadishu ridiculed U.S. attempts to capture him; hailed each failed raid as a victory; accused the UN and the United States of neocolonialism and of violating Islamic law. This campaign was reflected in TV images of angry Somalis burning tires, chanting anti-UN slogans, and ultimately, dragging an American serviceman's body through the streets of Mogadishu. These images came to symbolize, in the mind of many Americans, the Somalis' rejection of outside intervention. This perception did not match reality. According to Mark Walsh, a U.S. Army Peacekeeping Institute expert who was detailed to be the Special Representative for the Secretary General in the Southern Region (Kismayo), most Somalis supported the UN intervention because they longed for peace. Even after the events surrounding "Blackhawk Down," polling suggested that most Americans favored remaining if Somalis wanted the international community to remain.[11] Congress and most pundits, believing otherwise, pushed hard for a U.S. withdrawal.

In no small part, the United States and the rest of the international community fostered Aideed's approach, forgetting that Aideed and many of his compatriots were far from unsophisticated with wide-ranging educational and professional backgrounds. Aideed and his circle proved savvier about Western societies than the West did about Somalia.

THE UNITED NATIONS IN BOSNIA

Bosnia was first shattered by words, then by guns.[12] Dusan Basic

Indeed, propaganda and control of the airwaves played a major role in the forging the conflict surrounding the breakup of Yugoslavia and justifying the various factions' strategies. The first step in this process occurred in Serbia after Slobodan Miloševiæ arose to power in 1987. One of his first measures was to replace the leadership of TV Belgrade with supporters of the "new Serb nationalism." In the process, he fired several hundred journalists who did not subscribe to his agenda. Following this house cleaning, Serbian television began promoting the official national myth of the Serbian people's eternal martyrdom. Throughout the 1990s, Serbian media reinforced the memory of crimes committed (or supposedly committed) by other communities against the Serb people. The propaganda demonized

Kosovar Albanians, Croats, and Bosniaks and paved the way for justifying the war aims and the atrocities committed along the way. The second step was taken when the war broke out, first in Croatia (June 1991) and in Bosnia (April 1992). In each case, nationalistic forces took control of the existing media as well as created new ones to broadcast their messages. They instituted censorship and limited or eradicated the few existing independent voices.

Through this process, in all camps, the media became loyal instruments of the nationalistic parties' policies of war and ethnic purification. People's horizons shrank as the media portrayed reality in simplistic terms, demonized other ethnic groups (by exaggerating or inventing their crimes and responsibilities), and offered simplistic explanations for what was a complex and ambivalent reality. Such propaganda enabled the parties to mobilize their publics in favor of war goals and justified whatever means they used to accomplish those goals. Tadeusz Mazowiecki, former Polish Prime Minister and special rapporteur for the UN on the media situation in Bosnia concluded: "From the war's outbreak, the media in former-Yugoslavia mostly published and broadcast national discourses, attacks and other general insults directed against other ethnic groups. It is not surprising that this led directly to horrible atrocities on the battlefield and throughout the territory."[13]

By and large, the UN proved unable (and, mainly, unwilling to try) to manage the information environment. As the mission did not stop the fighting and left ethnic purification intact, the UN Protection Force (UNPROFOR) quickly lost credibility with the international press and came under increasing fire. The international press did not fault the UN for not delivering humanitarian aid, but faulted it for letting aggression and ethnic cleansing go unpunished and most of all, unstopped. As ethnic cleansing, concentration camps, and mass murders were uncovered in the summer of 1992, the press became increasingly hostile toward the UN leadership and demanded greater action from the peacekeepers. UN-press relationship became increasingly contentious and hostile, to the point where the UN was unable to communicate its agenda. In fact, it is not unreasonable to say that the UN, in essence, washed its hands and gave up on communication. The hostility came to the point where the UN would send a Spanish spokesman to the daily briefing who was just competent enough in English to read a prepared statement, but not competent enough to answer questions.

In addition, as the UN mission did not satisfy any of the parties to the conflict, it quickly became the target of the factions' propaganda campaigns. All sides accused the UNPROFOR of favoring the other side. For example, the Croats accused the Russians of supporting the Chechniks (derogatory term used to describe the Serbs) while the Bosniaks accused the French of pro-Serb bias. Meanwhile the factions manufactured incidents for the benefit of international cameras as a way to dramatize the suffering of their own. Gen. Lewis McKenzie of the Canadian army, the first UNPROFOR commander, argued that a durable cease-fire in Sarajevo could not occur if the parties did not stop firing on their own people for the benefit of CNN. Many Canadian and French officers who have served in Bosnia are convinced that hostilities increased each time CNN arrived in Sarajevo.[14]

To hobble UNPROFOR, the Bosniak government sought to discredit its commanders. In 1992, they accused General McKenzie of having raped four Muslim girls. Although it was baseless accusation, it so damaged General McKenzie's credibility that he asked to be relieved of his duty before the term of his mission had expired. He later explained that he felt these accusations had weakened his and the UN's authority to the point where the safety of his troops was no longer guaranteed. The Bosniak government repeated the same propaganda against Gen. Michael Rose of the British army. All these attacks diminished the credibility of UNPROFOR, constrained its ability to conduct its mission, and incapacitated efforts to strengthen its mandate.

The parties continued these efforts after NATO's deployment in December 1995. Similar to Aideed's propaganda in Somalia, pro-Bosniak outlets accused the French military of dumping nuclear waste on Mount Igman in part due to a Bosniak perception that the French were pro-Serbian. The Serbs regularly used disinformation tactics as well. For example, pro-Serb outlets reported toxic contamination near sites chosen by the UNHCR for minority returns in an apparent attempt to discourage those minorities from desiring to return.

THE CAMPAIGN AGAINST THE TALIBAN AND AL QAEDA (2001–ON GOING)

Taliban perception management sought to erode the international legitimacy of U.S. operations. Even though basic, the Taliban propaganda created headaches in the White House and the Pentagon. Exaggerated claims of civilian casualties beamed around the world unchecked while the Pentagon took days, if not weeks, to investigate facts—and there were few U.S. reports on these collateral damage incidents. Official explanations regarding the careful targeting process to minimize collateral damage were lost in the noise. Failures to share intelligence information with public affairs hampered effective responses. The world heard of attacks on Red Cross facilities even when the U.S. military knew that the struck ICRC warehouse was actually a military warehouse housing military vehicles and equipment with relief supplies under the same roof–a clear military target under international law.[15] Unsubstantiated claims of chemical weapons' use testified to the barbarity of the infidels in the Muslim world. Additional proof, if any was needed, that the United States was engaging in a war of civilization against Islam.

Initial U.S. rhetoric reinforced this perception. President Bush's reference to a worldwide crusade against terror and the code-name Operation Infinite Justice both created strong reactions in the Arab world. Vocabulary suggesting the Christian crusades against the infidels (mainly Muslim) and suggesting that the Americans appropriated to themselves divine power (as Allah is viewed in Islam, as God in Christianity, as the ultimate judge) was not designed to build Arab support for U.S. operations.

With the ever-widening reach of information technology, the Taliban had worldwide impact with even greater ease than Aideed. Key to this reach was Al Jazeera's presence in Kabul and the 4:30 follies in Islamabad.

■　Al Jazeera is the first independent twenty-four-hour news media in Arabic. It is based out of Qatar and is modeled after the BBC.[16] Its basic talk show and news format, much like Western media, is to present two sides for most issues (except Palestinian issues). Al Jazeera has captured much of the Arab audience in the Middle East and around the world. Al Jazeera has been the favored outlet for bin Laden videotapes.

■　For the first two months of Operation Enduring Freedom, Mullah Abdul Salam Zaeef, the Taliban's representative in Pakistan, was the most effective spokesman for the Taliban cause worldwide. Every afternoon at 4:30, he held a briefing with worldwide coverage that gave visibility to the Taliban position. Under great pressure from the United States, Pakistan cut off Zaeef in early November 2001.

Initially, the U.S. government seemed almost paralyzed by Taliban propaganda, with an incomplete and inadequate response to these outlets. The USG initially asked U.S. media outlets not to broadcast bin Laden's tapes and interviews since, according to Vice President Dick Cheney, al Qaeda could be using these tapes to send coded messages. Not having these tapes on U.S. media did not, however, greatly limit their reach. This explanation seems suspect as (1) these tapes were available via cable, satellite, and the Internet and (2) multiple other routes existed (Internet) to more effectively send "coded messages." This request seems more a ham-fisted PM effort to keep bin Laden off American television screens. Not just ham-fisted, also misguided. The American public would have, however, likely reacted with anger at bin Laden videos and they would have likely engendered ever-greater public support for the campaign against al Qaeda.

In terms of Al Jazeera and its impact in the Arab world, senior administration officials started giving high-profile interviews after weeks of essentially unanswered propaganda. To compete with Mullah Abdul Salam Zaeef, with urging by the British, the USG set up a Coalition Information Center (CIC) in Islamabad, Pakistan. The Pakistan CIC was supposed to release *all* information necessary to answer the Taliban propaganda. (This was a priori an impossible task, as coalition forces did not, for example, have on-scene knowledge of what civilian casualties may or may not have been in every incident.) For the initial months of the conflict, the State Department did not significantly expand Voice of America (VOA) programs and outreach throughout the Middle East. Throughout the operation, despite these efforts, Arabs remained skeptical (at best) of U.S. motivations and much of the Arab media treated USG assertions with disdain.

The Arab world does not share U.S. perspectives as anti-U.S. rhetoric dominates rumor intelligence (RUMINT) on the Arab streets. The following remain common assertions in Middle Eastern websites and news media:

■　The USG manufactured evidence against bin Laden, who was not involved in 9/11;

■　The Israelis (the Mossad) were behind the attack;[17]

- Clearly Arabs were not behind the attack, as no cell phone conversations from the hijacked planes mentioned the presence of Arabs; and,
- The United States is using 9/11 as an opportunity to extend its imperial reach.

These (and countless other) rumors, which have bounced around the Internet since 9/11, fuel and are fueled by a fundamental anti-U.S. sentiment. Core beliefs are that the United States:

- Is arrogant, hypocritical, and imperialist;
- Tolerates double-standards where they fits its agenda (does not tolerate al Qaeda, but tolerates and supports Israeli terrorism; and, waxes eloquently about democracy while propping up regimes like Saudi Arabia);
- Is a corrupt and decadent society run by Jews; and,
- Is waging a war of civilization against Islam.

While the realities and news coverage of the Israeli-Palestinian conflict hinder U.S. PM efforts, they are more fundamentally hobbled by the difficulty in conceiving the wide bridge in worldviews that exists between ourselves and those we are trying to influence.

OPERATION IRAQI FREEDOM (2003–ONGOING)

Operation Iraqi Freedom has also illustrated the difficulties in conducting effective perception management operations in the Middle East. Prior to the war, American efforts to convince the world that Saddam Hussein was a grave threat to U.S. and Western security were weak and unconvincing. Most of the world's publics did not see the necessity to go to war and demonstrated vocally against such prospect.[18] Many governments who chose to align with Washington did so not necessarily because of a shared vision on the problem, but to show solidarity with the United States. In Iraq, many believed that the United States was less interested in liberating the country from Saddam Hussein than in extending its control over the country and its resources.

The wartime PSYOP effort had mixed successes. PSYOP efforts helped deter the regular Iraqi army to engage in the fight. However, this effort failed to get the Iraqi population (in particular the Shi'a community in the South) to raise and take up arms against the regime (despite a closely coordinated campaign at the operational and strategic levels). On the public affairs side, DoD managed the American media to focus mostly on the U.S. military, but failed to do the same with international and Arab media. The latter increasingly focused its attention on civilian casualties, collateral damage, and excessive use of force. Some Middle Eastern journalists felt the U.S. media acted as a mouthpiece for the U.S. government and resentment toward America's free press surged in the Middle East when CNN and other outlets avoided coverage of casualty issues.[19] Meanwhile, as the controversy over the weapons of mass destruction balloons in the United States and Great Britain, many Europeans feel their anti-war position has been vindicated.

Post-war efforts have also been plagued by problems and difficulties to promote the U.S. agenda. The U.S. military has undoubtedly been able to encourage friendly behavior toward U.S. goals in several respects. Commanders have gained increased cooperation from the locals in the hunt for the fifty-five wanted former leaders. The coalition is also able to proceed, albeit with some difficulties, in creating new police forces and a new army. Civil affairs and PSYOP troops have been instrumental in explaining America's point of view and agenda to local leaders to entice their cooperation.

However, efforts to shape the political transition have been far less successful so far. The Coalition Provisional Authority's (CPA) first two transition plans have been scraped for lack of support by large segments of the population and their leaders. The CPA's efforts to secure basic law principles such as the separation of church and state, the recognition of minorities' and women's rights have met great resistance. Similarly, U.S. efforts to introduce Western-style media in Iraq have not been very successful. The decay of public diplomacy overall since the beginning of the 1990s has marginalized American voices in the Middle East.[20] Not surprisingly, the programs set up since OIF have not been able to find a large audience and rival with Arab networks that take the lion's share of the viewership. As if those challenges were not enough, U.S. efforts have been hampered by a series of budgetary constraints and organizational problems that have limited America's outreach.

Tipping the Scale in Our Favor

Put in a polite method, the United States has been marginally effective in its perception management efforts across the Arab world since 11 September. Considering past performance in this arena highlights a requirement to seriously examine issues that impair information campaigns and determine possible improvements. The following paragraphs raise issues from the highest political leadership to the command environment for military operations in the field. Absent addressing, these issues will continue to hobble U.S. PM efforts.

STRATEGIC-LEVEL ISSUES

Definitions: PA and PM

Perception management is not a well-defined concept. In the U.S. military, psychological operations is the only element defined as a PM tool. By law, PSYOP must be directed only at foreign audiences. By law, the public affairs role is to inform the American public, Congress, and news media. PA is not supposed to engage activities designed to shape attitudes. However, this is a distinction that loses merit each passing day. Public affairs, like any public communication, shapes the information environment. On the other hand, PA professionals are concerned that a close relationship with PM activities could undermine the credibility required for effective communication. Thus many view as inappropriate suggestions that public affairs participate in perception management.

As a result, PA is often marginalized in perception management/IO campaigns, in spite of the fact that PA deals with one of the major attitude-shaping elements in the information environment: the news media. This can weaken PM campaign efforts. For example, public affairs officers (PAOs) do not typically provide input about potential media fallout of specific targeting—in no small part because PAOs frequently are not involved in strike planning. PAOs often are far enough out of the loop so as not to have knowledge of specific situations or the context of operations. This undercuts PA effectiveness and, by extension, PM effectiveness as well.

Who Should Own PM?
The rationale for PM to belong to IO is twofold. There is a human dimension to information and PM deals with it. Prior to IO's emergence as a field, PSYOP was part of command and control warfare (C2W), which is the core IO competency. On the other hand, IO includes many specialties that have little to do with each other. Practically, the lead agency given the responsibility for IO will have a tremendous impact on IO definition and practice. As SPACECOM owns IO, it seems likely that PM will be on the backburner. When it comes to peace or post-conflict operations PM, how many senior leaders at SPACECOM have significant on-the-ground experience working with international organizations in a post-conflict situation? No matter the intentions and skills of the SPACEOM leadership, it is hard to imagine that this lack of senior-level experience will foster robust and innovative approaches to developing more effective PM in confused environments from Afghanistan to Baghdad and Kosovo to the former Zaire.

Strategic Leadership
Effective communication builds on a strategic-level plan that outlines the rationale, objectives, and means of the operation. In the USG, no single body is in charge of this short of the president. Once the president has articulated a vision (which doesn't always occur), the government must disseminate that vision quickly and accurately to all agencies and command elements involved in perception management. All too often, this process is reduced to mid-level bureaucrats using the president's and cabinet officials' speeches to sketch out PSYOP leaflets, radio programs, and/or public affairs guidelines. This inadequate and haphazard process needs strengthening to achieve a closer relationship between the national leadership and those in charge of spreading the USG's message and vision throughout the world.

Coordination, Cooperation, and Communication (C3) Critical
It seems clearly understood by most players that PM requires improved C3. In the USG, various agencies work with bits and pieces of the PM puzzle. Amid all its other tasking, the NSC staff holds a coordination responsibility. Effective PM requires effective multi-agency coordination. Until 1997, this coordination

occurred on an ad hoc basis, whenever a crisis erupted and the White House perceived an emergent need for a coordinated message. The ad hoc approach sometimes had great success . . . sometimes it didn't. In April 1999, President Clinton signed Presidential Decision Directive (PDD) 68 designed to establish a standing coordination mechanism for public diplomacy and psychological operations. PDD 68 implementation has been plagued with bureaucratic rivalries and minimal high-level involvement. The public fiasco of the Office of Strategic Influence (OSI) resulted from an effort to create improved C3 within the Department of Defense.[21] OSI failed, in no small part, due to the just-mentioned bureaucratic rivalries. One part of DoD sought to discredit OSI based on fundamental and unresolved disagreements over the most effective path toward influencing perceptions worldwide. As an aside, an indication that OSI was not necessarily ready for prime time came in the awkward and hesitant explanations of OSI that followed leaks about its existence. The U.S. government, and its coalition partners, requires a coherent and effective mechanism for coordinating and communicating messages beyond ad hoc CICs.

Catering to Different Audiences

The U.S. government's leadership has yet to realize the effects of media globalization, as it speaks primarily to the U.S. domestic audience. Catering principally to the national audience may create inadvertent messages worldwide, as comments targeted at U.S. citizens are seen around the world. The Global War on Terrorism has provided a rash of examples of this process at work. Seeking to foster America's support for a long-lasting war against al Qaeda and the Taliban, the political leadership sought to seize the moral high ground. The government, starting with President Bush, resorted to crude good-against-evil campaign imagery, full of religious symbols and morality (evil-doers, crusade against evil, axis of evil, Operation Infinite Justice). This approach and vocabulary reinforced perceptions in the Middle East that America was going to war with Islam. Allies (especially in Europe) saw this as an immoderate and unsubtle approach, likely to undermine effectiveness in the Arab world. Catering to the U.S. audience backfired on the international level—not for the first time and, sadly, such catering has continued since. In part, there is a greater sophistication in the perception management of the American public than of the world or in the operational arena. The expertise and resources of those who have political motivations—for example, the White House—for decades has outmatched the military and public diplomacy resources for perception management. The president's political advisers, concerned about the next election, are far more likely to affect nuance in a State of the Union address than some regional specialist in a State Department bureau or regional combatant commander's staff. This imbalance in access, power, and resources will be hard to overcome unless the most senior leadership recognizes that the consequences of this imbalance are too serious for the nation's security amid the Global War on Terrorism.

PM as EBO

A key arena for DoD thinking and debate revolves around effects-based operations (EBO). The key question: How can military tools best assist the nation to achieve its fundamental objectives? Across the conflict spectrum, this means an even greater focus on effecting the will and perception of the opponent. To do this effectively requires a level of understanding of adversaries, potential adversaries, and partners that does not seem extant at this time. The history of the use of force related to Iraq (Saddam Hussein) and Serbia (Slobodan Miloševiæ) over the past decade highlights how limited understanding can be in knowing what will effect opponents.

Even as globalization blurs certain differences between cultures and societies (or seems to), the United States cannot necessarily influence friends and deter foes effectively through projecting U.S. culture and values on others when and where they are not welcome. In many places from Iran to France, a backlash to elements of American culture is a reality. Globalization will not lessen this trend, with the backlash differing tremendously across the world community. Changing the traditional approach requires that the United States: improve understanding of how various tools of national power interact in a globalized world; develop a more reliable set of tools for analyzing adversaries; and construct a national (and international) decision-making and execution structure that facilitates using these tools to achieve desired results.

OPERATIONAL-LEVEL ISSUES

Defining the Global Information Environment

Just as intelligence support is required to put bombs on the right target, intelligence support is critical for effective PM efforts. The following are critical questions requiring examination prior to launching PM activities.

- How does information circulate in target audiences?
- What local means of communication are available to enhance message credibility?
- What fault-lines exist between your and your target audiences' belief-systems?

Examination of the experience in recent decades suggests that these and other critical questions are rarely well understood when PM efforts commence.

PM Requires Command Attention and Close Coordination

Command attention fosters effective PM. Commanders with effective PM efforts typically hold daily meetings with key actors, including PA, PSYOP, intelligence, and operations.

Coordinate Internally

Fully effective information activities are tied to operations. Close integration with other operational staffs (in particular the 3 shop) fosters more effective preparation for and response to contingencies. This includes regular meetings of PA, PSYOP, and civil information with operational and intelligence staffs to receive inputs on the information campaign and provide input to the targeting process.

Coordinate Externally
Coordination also needs to exist between coalition military partners and the numerous civilian agencies likely to be engaged in the battlespace. These include national agencies (such as the U.S. Agency for International Development); international organizations (UN or EU agencies); and non-governmental organizations (such as CARE). The military is rarely the sole actor in an international operation, especially during peace operations. PM C3 with these organizations will enhance mission effectiveness and speed mission achievement.

A Concluding Thought
Perception management is a critical tool for the twenty-first-century warrior and peacekeeper—and for every stage between the two. As project TROY recognized more than fifty years ago,

> a rifle becomes meaningless as a mere component of military power unless combined with artillery, tanks, aircraft, and all the other necessary weapons. Like a rifle, an information program becomes a significant instrument in the achievement of our national objectives only when designed as one component in a political weapon system. Political . . . warfare should be organized like any other form of warfare, with special weapons, strategy, tactics, logistics, and training. [22]

As the situation stands today, PM structure, concepts, and approaches combined with media developments in the democratic world leave this field as a key asymmetric vulnerability. Change is required across the process to turn this to Western advantage. As indicated above, change is required within the highest political leadership (such as realizing the problems caused by attempting to differentiate between domestic and international messages) and within military tactical efforts (which need to inculcate the concept that PM is now a critical tool for all military operations and that every member of the military and associated organizations is, inherently, part of that effort). In the GWOT, PM will be a critical tool for eventual victory. Failure in this battlespace will undermine the potential for achieving that victory.

CHAPTER 4

Information Operations in the Global War on Terror: Lessons Learned From Operations in Afghanistan and Iraq

Zachary P. Hubbard, Lieutenant Colonel, U.S. Army (retired), Former IO Division Head, Joint Forces Staff College.

On 11 September 2001, nineteen airplane hijackers, using four hijacked commercial airliners, conducted a series of attacks against the United States that radically altered the American worldview. Most of the perpetrators were Saudi Arabian nationals and followers of Saudi dissident Osama Bin Laden. We may never know their complete target list. However, the two targets successfully hit were carefully chosen symbols of America's national power. The World Trade Center (WTC) complex symbolized America's global economic prowess. The Pentagon is recognized worldwide as the symbol of America's military might. Some believe the U.S. Capitol Building, a symbolic center of the U.S. political system, was the intended third target; it was spared, apparently due to the heroic efforts of individual citizens who took charge of events on United Airlines Flight 93 after learning of the attacks on the WTC and the Pentagon. The plane crashed in an empty field in Pennsylvania, missing its target but killing all onboard. Only twenty-six days later, a U.S.-led international Coalition that would grow to over fifty nations invaded Afghanistan in the initial phase of what President George W. Bush has termed the Global War on Terrorism (GWOT).[1]

The invasion was a direct response to the 11 September attacks. Operation Enduring Freedom (OEF) aimed to eliminate bin Laden's al Qaeda terrorist group at its primary operating base, Afghanistan. At the time, Afghanistan was a country ruled by the radical Islamist Taliban regime and the host to numerous training camps for terrorists from around the world. Its torturous terrain, paucity of economic infrastructure, and reputation for having defeated the invading Soviet Army in a bloody, protracted guerrilla war promised to make Afghanistan a difficult battlefield.

In March 2003, twelve years after winning Operation Desert Storm, a U. S.-led Coalition began the second major phase of the GWOT, invading Iraq. Iraqi dictator Saddam Hussein had long stood accused of aiding and abetting terrorists and had defied the United Nations since the end of Desert Storm by refusing to comply with the surrender terms. While the U.S. could make no direct connection between the Iraqi regime and the attacks of 11 September, Iraq's long history of harboring terrorists and its pursuit of weapons of mass destruction were deemed sufficient by the administration to warrant a regime change. The DoD christened the second Gulf War, Operation Iraqi Freedom (OIF).

This chapter examines the United States' employment of IO in the initial combat and early follow-on operations in OEF and OIF. It also discusses IO lessons learned during major combat operations and during the subsequent periods of stability operations and nation building. While the focus of the discussion is IO in the DoD, I also address the strategic communications and public diplomacy conducted by the Unites States Government (USG), because DoD IO must complement these programs. Strategic communications are the media themes and messages originating from the executive branch of the USG, particularly the White House and Department of State. Public diplomacy consists of U.S. diplomatic communications aimed at the citizenry of a foreign country, rather than towards the government leadership, as is the case with traditional diplomacy.

Looking back, in Operation Desert Storm the art and science of Command and Control warfare was applied to near perfection. Joint Command and Control Warfare doctrine is established in Joint Publication (JP) 3-13.1. It defines C2W as, "The integrated use of operations security, military deception, psychological operations, electronic warfare, and physical destruction, mutually supported by intelligence, to deny information to, influence, degrade, or destroy adversary command and control capabilities, while protecting friendly command and control capabilities against such actions." Command and Control warfare targets an adversary's Command and Control (C2) systems. Military commanders use C2 systems to coordinate and synchronize virtually every aspect of their planning and operations. Applying C2W, the Allied Coalition in Desert Storm systematically bombarded Iraq's frontline troops with psychological operations (PSYOP); crippled Iraq's integrated air defenses; blinded its target acquisition; shut down its propaganda machine; and totally disrupted military communications from the strategic to the tactical levels. These actions were supported by the most massive military deception operation since the Normandy invasion, which featured General Norman Schwarzkopf's famous "left hook," an operational maneuver that projected massed armor deep into Iraq to destroy Saddam Hussein's vaunted Republican Guard. Iraq's military, psychologically exhausted and rendered electronically deafened and blinded by C2W, were subsequently defeated in detail by relentless aerial attacks combined with a lightning-swift ground campaign. The image of defeated

Iraqi forces surrendering by the thousands remains vivid today. C2W was the genesis of the joint IO doctrine employed by the U.S. military today.

Despite the U.S. military's enthusiasm for IO and having had over six years to exercise its joint IO doctrine, at the time of this writing the DoD and military services have only recently agreed on a common definition of IO. DoD Information Operations are,

> the integrated employment of the core capabilities of electronic warfare, computer network operations, psychological operations, military deception, and operations security, in concert with specified supporting and related capabilities to influence, disrupt, corrupt, or usurp adversary human and automated decision-making while protecting our own.[2]

The emerging IO doctrine specifies public affairs (PA), civil military operations (CMO), and military support to public diplomacy as functions specifically related to IO.

At the time of this writing, the Department of Defense IO doctrine in Joint Publication 3-13 is undergoing a major revision. My discussion in this chapter divides IO into three categories, based upon the recently revised IO doctrinal framework of the U.S. Air Force, which I find a convenient and logical way to "package" IO discussions: influence operations, electronic warfare operations, and network warfare operations.

Influence operations are the integrated planning and employment of military capabilities to achieve desired effects across the cognitive battlespace in support of operational objectives. Psychological operations (PSYOP), military deception (MD), operations security (OPSEC), counterintelligence (CI), and public affairs (PA) are elements of influence operations. Electronic warfare (EW) operations are the integrated planning and employment of military capabilities to achieve desired effects across the electromagnetic battlespace in support of operational objectives. Electronic attack (EA), electronic protection (EP), and EW support are the operational elements of EW operations. Network warfare (NW) operations are the integrated planning and employment of military capabilities to achieve desired effects across the digital battlespace in support of operational objectives. Network attack and network defense are operational elements of NW operations. For simplicity I have grouped the discussion of strategic communications, public diplomacy, and civil military operations with the discussion of influence operations.

The reader will note that while the discussions of major combat operations include observations on influence operations, electronic combat operations, and network combat operations, the discussion of stability operations discusses only influence operations. This is due to a paucity of open-source materials on the latter two areas. This should come as no surprise, as the necessity to win hearts and minds during stability operations significantly restricts the use of electronic combat operations and network combat operations. The DoD IO for the remainder of OEF and OIF will focus mainly on influence operations. It would therefore be wise to focus on analyzing influence operations lessons learned to ensure that mistakes in this critical area of IO are not repeated. Media operations, PSYOP, and civil military operations will play a key role in the success or failure of these operations. All information used in writing this chapter came from open sources.

The circumstances surround the beginnings of OEF and OIF were considerably different. This affected the employment of IO during each. The employment of IO in each operation also reflects the complexity, or lack thereof, of the communications infrastructures, both technical and human, in each area of operation. IO in OEF has been dominated by influence operations from the beginning. IO during major combat operations in OIF was a more balanced mix of influence operations, EW operations, and NW operations. IO during stability operations in OIF has been almost exclusively influence operations. With the targets of 11 September still smoldering, the U.S. enjoyed considerable global support for invading Afghanistan. By the beginning of OIF, a group of former allies, led by France and Germany, had blocked support in the United Nations Security Council for a U.S. invasion of Iraq.

OEF was begun with virtually "no-notice," meaning the Pentagon had only a minimal amount of time in the wake of 11 September to prepare the information battlespace prior to the initiation of hostilities. Consequently, IO suffered from a lack of sufficient planning and preparatory time. Afghanistan lacked a robust national communications infrastructure, making it impossible to inform the Afghan citizens of Coalition intentions using an information campaign conducted through the international media. The Coalition did not have sufficient time to employ a lengthy PSYOP campaign before beginning hostilities, as would normally have been desired. The high rate of illiteracy amongst the Afghan population increased the difficulty of producing effective, printed PSYOP products.

> When the Taliban took control of the government in Afghanistan, women were no longer allowed to hold regular jobs. This directive by the Taliban reduced the number of medical professionals and more than half of the country's schoolteachers. Furthermore, by restricting what and who (no women) could be taught in school, the Taliban further limited the education of the people. This decision coupled with the fact that many of Afghanistan's middle class left years ago during the conflict with the Soviets have resulted in a country of discontented youth, undereducated, unemployed, restless, resentful of their fate.[3]

The Afghan population exists in a loose confederation of clans led by feudal warlords. In the beginning of the conflict, most did not even understand why the Americans had invaded. "Information warfare experts look for what they call 'the voilá moment.' In Afghanistan, the biggest lesson we learned in our tactical information operations— the radio and TV broadcasts—was the importance in explaining, 'Why are we here?' a senior American military officer said. 'The majority of Afghanis did not know that 11 September occurred. They didn't even know of our great tragedy.'"⁴ The "voilá moment" in OEF wouldn't come until the Afghan population understood why the Americans and their Coalition partners had come to fight the Taliban. IO was the primary means of providing an explanation.

The situation at the beginning of OIF was considerably different. For all practical purposes, the United States and the United Kingdom had been at war with Iraq since 1991, flying daily combat missions to enforce the no-fly zones over Iraq in Operations Southern Watch and Northern Watch. Iraq's well-educated, fairly sophisticated population boasted high literacy, a taste for Western media and entertainment, and enjoyed a robust national communications infrastructure. This included widespread access to global radio and television broadcasts via satellite. Women were well integrated into the machinery of the Iraqi government and society. Iraq was a far more fertile ground for waging IO than was Afghanistan. The situation in each country dictated the very different manners in which Coalition forces employed IO on the two main, foreign playing fields of the GWOT.

Influence Operations during Major Combat Operations in OEF

For countries that could never hope to match the United States militarily, IO has leveled the playing field somewhat. Even when a country lacks good communications infrastructure, like Afghanistan, the international media is readily available to transmit propaganda, whether wittingly or unwittingly. It is clear that Afghanistan's Taliban leadership manipulated the media to spread propaganda aimed at turning world opinion against Coalition operations. The tactics used had been seen before in the Balkans: intentionally damaging mosques to make it look as if they had been damaged by Coalition bombing; taking the press on well staged hospital tours to see supposed victims of Coalition atrocities; and placing military vehicles in locations near mosques or other public buildings in the hope that collateral effects of an attack on the vehicle might damage the mosque or building, leaving incriminating "proof" of war crimes.⁵ DoD officials held several press briefings during OEF and OIF, specifically for the purpose of exposing enemy denial and deception (D&D) methodology and explaining in detail how they could prove that the purported evidence was faked. Most of the evidence came from comparing overhead imagery of the target areas before and after the attacks. This approach was fairly effective in neutralizing some of the enemy propaganda that might have otherwise required much greater effort to overcome.

The influence of the Qatar-based Al Jazeera broadcasting network, which has been called the "Arab CNN," cannot be overstated. Al Jazeera beams its message across the Middle East and is reported to reach thirty-five to forty million viewers.⁶ It stands accused by some of being a mouthpiece for Osama bin Laden, periodically airing recorded messages from bin Laden that may be used to pass cryptic messages to his followers. Regardless of its motives, Al Jazeera is a market competitor for U.S. PSYOP and public affairs broadcasts in the Middle East.

The first such bin Laden tape broadcast by Al Jazeera demonstrated just how much confusion remains in the USG regarding IO, especially when it comes to harmonizing media operations. "When the first bin Laden tape was broadcast on Al Jazeera television, followed by governmental requests to commercial stations not to rebroadcast for fear it may contain coded messages to terrorist 'sleepers,' Voice of America ignored the request."[7] For some reason, the U.S. government-operated Voice of America failed to comply with the request, even as some commercial media outlets suppressed it. The fact that the USG asked commercial media to suppress the tapes was used as ammunition by the enemy to accuse the "free speech loving" Americans of hypocrisy.

Al Jazeera was hard at work early during OEF, broadcasting messages from Osama bin Laden even as U.S. Army troops searched for him amongst the caves in the mountainous terrain along the border between Afghanistan and Pakistan, where he was believed to be hiding. For a while, Al Jazeera also aired daily interviews with Afghanistan's ambassador to Pakistan. The ambassador, a member of the Taliban regime, used these interviews to spread misinformation about alleged civilian casualties resulting from American combat operations in Afghanistan. This one-sided coverage proved troublesome for the U.S. leadership, who had difficulty deflecting international criticism of their operations that grew steadily with the daily interviews of the ambassador.

The U.S. made limited use of television in an attempt to explain the 11 September attacks to the Afghan population. Few people in Afghanistan own televisions, so broadcasting a television message would have been of little value. Instead the Americans produced a three-minute video explaining the attacks and promoting support of the government of President Hamid Karzai. The video was then shown around the country through a series of visits with local officials and citizens by U.S. PSYOP troops.[8] Some individuals who saw the video obviously understood. Others, some who had never even seen television before and who were unfamiliar with the images of a big city and skyscrapers, seemed not to understand. The overall effectiveness of this methodology cannot be accurately assessed, but under the circumstances it was probably the best that could be accomplished using televised media. One of the key requirements for effective PSYOP is selecting the proper media and products to suit the target audience. Television was not the media of choice for broadcasting PSYOP messages in Afghanistan.

As in most recent conflicts, PSYOP leaflets were a key element of U.S. influence operations in OEF. On 8 November, only a month into the operation, the Pentagon announced that sixteen million PSYOP leaflets had been dropped on Afghanistan.[9] Because of the Afghan population's high illiteracy rate, many leaflets were cartoon style, with little or no writing. The challenge of developing effective leaflets of this type is that different people will interpret them in different ways. This led to at least one situation with potentially deadly results. U.S. aircraft dropped humanitarian daily rations (HDR) in many areas of Afghanistan. The HDR, packaged in bright yellow plastic, are prepared in accordance with the strict dietary requirements of the Koran. Unfortunately, the HDR packets are similar in size and color to the sub-munitions in certain cluster bombs. The aircraft dropping HDR also dropped cartoon-style leaflets showing an Afghan-looking family eating the rations. When someone realized that the leaflets might lead some to pick up the yellow sub-munitions, the Pentagon had to scramble to correct the problem. They quickly produced a new leaflet warning the Afghan population not to pick up the sub-munitions that looked like HDR. Human rights groups had criticized the use of the cluster bombs because it was thought that children might be attracted to the bright yellow color of the sub-munitions. The Pentagon reported that the color of the humanitarian daily rations would be blue in the future.[10]

The Taliban regime published propaganda pamphlets aimed at discrediting Coalition forces. Offering monetary rewards for killing or capturing Westerners and threatening retaliation for cooperating with them, the pamphlets were slipped under people's doors during the night. This led to their being nick-named "night letters."[11] Similar pamphlets were also circulated among refugees in Afghanistan and Pakistan. The Taliban propaganda threatening violence against their own people stands in stark contrast to the U.S. PSYOP and media methodology. There was no apparent attempt by the Taliban to target U.S. or Coalition forces with propaganda, something that has proven fruitless in previous conflicts.

Electronic Combat Operations during Major Combat Operations in OEF

The simple description of EW operations in Afghanistan is that most of the limited Afghan communications architecture and air defenses were taken out by physical and/or electronic attack during the initial hours of the Coalition air attacks. The electronic warfare and air defense environment Coalition aircrews faced in Afghanistan can be described simply as low-threat or non-hostile. Due to this low-threat air defense environment, the U.S. Navy's EA-6B jamming aircraft, traditionally used for suppression of enemy air defenses (SEAD), were employed to "exploit new techniques to jam ground communications by working with the EW-130 and other electronic intelligence gathering aircraft."[12] The DoD displayed similar innovation during OIF, where the U.S. Air Force for the first time used the Compass Call electronic attack aircraft to broadcast PSYOP messages with its powerful transmitters.[13] Such innovation exemplifies the synergy that can be gained through combining the effects of IO capabilities.

The following paragraphs, while grouped in the discussion of electronic combat operations, could equally have been placed under the discussion of influence operations. This is typical of so many information operations. For example, bombing an enemy radio station to silence its propaganda broadcasts is considered a form of IO, according to current U.S. joint IO doctrine, because it is intended to thwart the enemy's influence operations.[14] Taking over an enemy frequency for your own uses would be considered electronic warfare in some circles. So whether the following examples constitute influence operations or EW operations, the lessons to be learned are the same.

Before OEF began, "the Taliban-run Voice of Sharia radio, broadcast from Kabul, filled the airwaves with religious discourse and official decrees. Their opponents, listeners were told, were 'evil and corrupt forces.' Today, that Taliban signal has turned to static, its transmitter destroyed by two cruise missiles."[15] Having taken out the transmitters used by the Taliban to spread propaganda about the Coalition forces, the U.S. military took over the main frequency used for Taliban radio broadcasts and used it for their own PSYOP broadcasts to the Afghan population. They also distributed some seven thousand radios to the Afghan people.[16] Because of the isolation of many Afghan villages, it was difficult to reach them with either radio or television broadcasts. This problem was addressed using typical American innovation. "So Special Forces troops made contact with local coffee-house managers, and offered them the same radio programs being broadcast from *Commando Solo* planes, but on compact discs to be played over a boom box for the patrons. The program gave birth to a new icon on the military's maps of Afghanistan: a tiny picture of a coffee mug to indicate the location of village businesses that agreed to play CD copies of the American radio programming."[17] The EW-130E *Commando Solo* is a cargo plane converted to conduct PSYOP broadcasts and is discussed later in this chapter.

The war in Afghanistan saw the Army's *Prophet*, an unmanned aerial vehicle (UAV), tested operationally for the first time. The initial feedback from users was extremely favorable. The *Prophet* is a multi-mission platform that provides the ground force commanders the capability to conduct tactical signals intelligence (SIGINT) operations, measurement and signatures intelligence (MASINT) operations, and electronic attack operations.[18] The *Prophet* employment is part of a trend of increasing use of UAVs in U.S. military operations. The multi-mission nature of UAVs make them prime candidates for answering some of the prime IO requirements of the future, including electronic attack, PSYOP broadcast, COMSEC monitoring, OPSEC evaluation, computer network attack, public affairs and civil affairs broadcast, and a host of other applications. As U.S. military technology evolves, the distinction between intelligence, surveillance, and reconnaissance (ISR) systems and IO systems will continue to blur.

Network Warfare Operations During Operation Enduring Freedom

While the details of any DoD NW operations employed during OEF remain classified, it is safe to assume that they had little if any impact on the operation. Afghanistan was a technologically unsophisticated adversary that posed no significant threat of computer network attack and likewise possessed little network infrastructure that would warrant a sophisticated CNA. Iraq would prove a much more fertile ground for NW Operations

Influence Operations during Major Combat Operations in OIF

During the build-up to OIF and since, OPSEC has been a constant concern for the USG. Secretary of Defense Rumsfeld issued new OPSEC guidance addressing the use of websites by the DoD on 14 January 2003, noting that an al Qaeda training manual recovered in Afghanistan states provided security-sensitive information about the United States and concluded that, "using public sources openly and without resorting to illegal means, it is possible to gather at least 80 percent of information about the enemy."[19] Before the 11 September attacks, it was possible to access a vast amount of operationally sensitive information from government, military, and commercial websites. Detailed maps of nuclear power generation plants, DoD military installations, and key public buildings such as the U.S. Capitol are just a few examples of information that was freely available to any terrorist with Internet access. It is now clear that all levels of government activities, the DoD, and key private organizations comprising the United States' critical economic infrastructures must have an OPSEC vetting process to review information before it is placed on a publicly accessible website.

The Bush administration received considerable criticism for suppressing the news media during OEF. As *New York Times* journalist Elizabeth Becker reported on the war in Afghanistan, the "military has imposed a tight lid . . . trying to walk the fine line of saying enough to reassure the public that the war is on target but keeping the news media at bay."[20] The so-called act of embedding journalists with Army and Marine Corps combat units during OIF was an answer, at least in part, to such criticism.

Embedded journalists enabled the world to see the "ground truth" in Iraq, which somewhat aided in neutralizing Iraq's propaganda efforts. However, the embedded journalist sometimes exacerbated the OPSEC problem. Reporter Geraldo Rivera was expelled from the 101st Airborne Division's zone of operation because of an OPSEC violation.

> Rivera violated the cardinal rule of war reporting by giving away crucial details of military plans during a Fox News broadcast from Iraq. During the broadcast, Rivera asked his photographer to aim the camera at the sand in front of him. He bent down and drew a map of Iraq in the sand showing the comparative location of Baghdad to his unit. Rivera even proceeded to draw diagrams of where the unit was heading next. If enemy forces were watching the news broadcast, they could have launched a preemptive strike against the whole Airborne Division.[21]

Rivera later admitted that he had made a mistake and apologized for the blunder. In the past the media has posed little immediate threat to combat forces. However, since the advent of global television and radio broadcast from the frontlines, OPSEC has become a major issue. The 1st Marine Division Commander, Maj. Gen. James M. Mattis characterized the embedded journalists concept as "a limited success."[22] His opinion of the journalists is not as positive as that expressed by Secretary of Defense Rumsfeld or U.S. Central Command (USCENTCOM) Commander General Tommy Franks.

OPSEC also became an issue on the home front. An egregious violation potentially put the family members of a B-1 bomber crew at risk during OIF. On 7

April 2003, a B-1 bomber crew conducting a mission over Iraq was diverted to a new target. The target was later revealed to be a building in which Saddam Hussein and some of his senior leadership was believed to be located. In a subsequent interview broadcast from the Pentagon, a number of crewmembers' full names, their commanding officer's name, their unit, and their home base location were identified.[23] Given the threat of terrorism in the U.S. today, this incident could have well put the family members of these individuals at risk. With today's powerful Internet search engines, given a name and location, it is often possible to obtain an individual's phone number, email address, street address, and a map to their place of residence in a matter of minutes. Identifying the unit, base, and full names of the B-1 crew during a media event was an OPSEC violation of the worst kind. Attacks on family members of deployed service men and women could have dire operational consequences for the U.S. military. Deputy Secretary of Defense Paul Wolfowitz issued a new OPSEC directive to the DoD on 6 June 2003, directing the military services to reassess their OPSEC programs by 1 October and providing new objectives for the assessment.

The U.S. military officially acknowledged the human-factor element of influence operations when it revealed that senior Iraqi military leaders were bribed into submission before the onset of ground combat in Iraq. General Tommy Franks, the USCENTCOM Commander, indicated that senior Iraqi officers accepted bribes for a promise not to engage Coalition forces. Consequently, Coalition forces met light resistance in many locations that might have otherwise been heavily defended. "I had letters from Iraqi generals saying: 'I now work for you,'" General Franks said.[24] This same type of influence was employed years earlier to encourage the peaceful departure into exile of Gen. Raul Cedras during Operation Uphold Democracy in Haiti.[25] Money, sex, and drugs remain three of the best, time-tested tools of influence in war and politics. The United States' adversaries employ these tools. Therefore, periodic counterintelligence training, emphasizing the methods employed by hostile intelligence services, is a critical tool for protecting critical information.

During OIF, President George W. Bush employed national strategic communications to directly address Iraqi generals in a nationally televised speech during which he made it clear that, "anyone ordering the use of weapons of mass destruction will be treated as a war criminal and likely will be executed."[26] As is often the case with influence operations, it is difficult to assess the effects of the president's warning. However, the large number of prisoners from OEF that were being held in a U.S. military prison in Guantanamo Bay, Cuba were ample evidence of the veracity of his words and may have contributed to the psychological impact and effectiveness of the warning. The DoD has yet to develop adequate measures of effectiveness (MOE) for evaluating influence operations and it may prove an impossible task, given that human behavior is so unpredictable.

As usual, PSYOP broadcasts from the air force's EC-130E *Commando Solo* aircraft were a key element of the OIF information campaign. One program broadcast by the EW-130E mimicked a popular Iraqi radio program, *Voice of Youth*.

"The American programs open with greetings in Arabic, followed by Euro-pop and 1980's American rock music—intended to appeal to younger Iraqi troops, perceived by officials as the ones most likely to lay down their arms. The broadcasts include traditional Iraqi folk music, so as not to alienate other listeners, and a news program in Arabic prepared by army psychological operations experts at Fort Bragg, NC. Then comes the official message: Any war is not against the Iraqi people, but is to disarm Mr. Hussein and end his government."[27]

The inability of *Commando Solo* to operate in areas of mid- to high-intensity air defense threat has long been a problem, as has the limited range of its television and radio broadcasts. A Defense Science Board study published in 2000 indicated that the small, aging fleet of EC-130E's is no longer capable of meeting the PSYOP needs of the combatant commanders.[28] To help remedy this, during OIF Coalition forces employed PSYOP broadcasts from ships operating in the Persian Gulf.[29] The U.S. Navy in recent years deployed a containerized PSYOP broadcast package that can be placed on and operated from the deck of a ship. The navy has also improved its ability to produce printed PSYOP products while underway, rather than having them delivered to their ships. As the military services continue to develop new concepts and requirements for operational platforms and systems, PSYOP broadcast, leaflet delivery, and new means of delivering PSYOP messages should be considered as a potential capability for each.

During OIF, the Al Jazeera network was once again at work. It received considerable criticism from American officials for broadcasting video of Americans captured by the Iraqis, including one that appeared to show that some of the prisoners had been summarily executed. Many American citizens cried foul and accused Al Jazeera of outright support of Saddam Hussein. The USG's response to Al Jazeera thus far has been rather clumsy. The government has yet to display a coordinated strategic communications strategy to deal with Al Jazeera and similar Middle Eastern media outlets. Until this happens the U.S. will continue to lose ground in global media confrontations.

OIF did have its lighter moments. Muhammed Saeed Al-Sahaf, the Iraqi Information Minister who is known today by his nickname "Baghdad Bob," became a celebrity through parodies of him on the David Letterman Show, the Tonight Show, and a website dedicated to his infamous railings (www.welovethe iraqiinformationminister.com). Al-Sahaf claimed in his now famous taped interviews that the Iraqi armed forces were slaughtering Coalition forces, even as Coalition tanks drove through the streets of Baghdad and Coalition forces occupied Saddam Hussein's palaces. The lesson for anyone who watched Baghdad Bob is that truth is the strongest weapon in PSYOP. Unlike the propaganda employed by most of its adversaries, U.S. PSYOP doctrine focuses on transmitting selected, truthful messages to specific human targets in order to influence their actions. Baghdad Bob is an excellent example of how not to conduct PSYOP. I triple guarantee it!

As in Operation Desert Storm before it, Iraqi denial and deception operations during OIF were effective. During Desert Storm the Coalition devoted considerable time and resources to locating and destroying Iraq's elusive mobile SCUD missile launchers, with no success. Some claim that Iraq's interaction with the United Nations during the twelve ensuing years after the first Gulf War was one massive deception operation. The certainty with which President Bush and Great Britain's Prime Minister Tony Blair accused Iraq of possessing weapons of mass destruction (WMD) and the inability of the Coalition forces now occupying Iraq to locate a single WMD suggest that Iraq may have pulled off one of the greatest deception operations in our time.

It is now clear that the Iraqis employed a number of innovative means to deny and deceive Coalition intelligence surveillance and reconnaissance during OIF. These included placing weapons facilities inside civilian neighborhoods, faking collateral damage for the sake of exploiting the media, and hiding weapons and munitions inside schools, mosques, and walls of buildings.[30] Scores of official documents have been recovered from private homes. Perhaps the most innovative D&D method was burying equipment and documents. The Army's 101st Airborne Division discovered eleven mobile laboratories, suspected of being biological agent production labs, buried near the town of Karbala."[31] "Officials said they have found tons of military equipment, including airplanes, buried beneath the sand, and they believe illegal weapons and the laboratories to make them will have been hidden in such a manner."[32] Finding buried materials in a country the size of California is like searching for a needle in a haystack. Much of the equipment and documents discovered so far was located as a result of tips provided by Iraqi citizens. Despite the U.S. military's sophisticated ISR capabilities, it is painfully clear that there is still no substitute for good human intelligence on the ground in an area of operations. However, sophisticated ISR proved useful for exposing Iraqi D&D techniques to the media, as was clearly demonstrated in a State Department media briefing on 11 October 2002. The briefing was dedicated solely to explain to the media the D&D methods used by Iraq during OIF.[33] D&D methods will only improve with time, meaning future ISR sensors will have to defeat increasingly sophisticated D&D methodology.

Electronic Warfare Operations during Major Combat Operations in OIF

Suppression of enemy air defenses (SEAD) remains critical to the U.S. Air Force's ability to accomplish its EW mission. One Coalition fixed-wing aircraft was lost to Iraqi air defenses during OIF. Another was downed by friendly fire. A software flaw in the army's Patriot Missile fire control system may have caused the friendly fire incident.[34] One of the key weapons in the SEAD effort is the AGM-88 HARM missile. The HARM is an anti-radiation missile employed against radar emitters. It is usually associated with the navy's EA-6B Prowler aircraft. Serbian air defense units in Kosovo learned that they could avoid the missiles by turning off the radar emitters after handing a target off to a missile launcher.[35] A recent software upgrade to the AGM-88 saw its first combat employment in OIF. The Block V and Block IIIa upgrades maintain their targets in memory after the targeted radar emitter is switched off. The upgrade also has the option to home in on emitters attempting to jam the Global Positioning System (GPS) receivers.[36] The AGM-88 upgrade proved effective in OIF. This may account for why so many Iraqi radar sites never turned on their radars at all, although it cannot be proven this was the cause. The improvements to the AGM-88 target memory are a good example of how the DOD and industry need to respond quickly when sophisticated adversaries develop D&D measures against U.S. technologies.

Coalition electronic deception was effective during the OIF air war. The U.S. Navy employed the tactical air launched decoy (TALD) system against Iraqi air defenses. The TALD is a small, air-launched glider that flies a preprogrammed course for about one hundred kilometers The Navy found that during night attacks in particular, Iraqi air defense guns and missiles often fired at the TALD decoys, reducing the risk to combat aircraft. As a result, the Navy has submitted orders for an improved TALD (ITALD) that has a small jet engine. As adversaries develop greater air defense technologies, it will be increasingly necessary for the DoD and industry to develop electronic D&D capabilities to protect U.S. aircraft.

By the time the Second Gulf War began, Iraqi forces had reconstituted their air defenses and could boast a robust, if somewhat dated, integrated air defense system (IADS). Before the war, U.S. military planners expressed concern over Iraq's acquisition of GPS jammers from Russia. This was due to the U.S. military's growing reliance on GPS navigation for precision weapons, particularly the GBU 31/32 Joint Direct Attack Munition (JDAM). Their concerns appear to have been unfounded. In a press conference on 25 March 2003 Maj. Gen. Victor Renuart, the USCENTCOM Operations Officer (J3), confirmed that the Iraqis had employed six GPS jammers during the previous two nights of operations and that all had been destroyed. He added that one of the Iraqi GPS jammers had even been destroyed by a GPS-guided precision munition.[37] It appears that DoD's heavy reliance on GPS may not have created as great a vulnerability as originally thought. However, greater numbers of GPS jammers employed by a more sophisticated opponent might have proven more effective. OIF simply did not provide enough data from which to draw firm conclusions on this issue.

Network Combat Operations during Major Combat Operations in OIF

While the details of sophisticated network attack operations are classified, the DoD did confirm that the U.S. employed a barrage of emails, faxes, and cell phone calls to numerous Iraqi leaders in an attempt to persuade them not to support Saddam Hussein.[38] As might have been expected, these technical NW operations were conducted in support of PSYOP. In the information age, email addresses, fax numbers, and cell phone numbers can be a valuable tool to an adversary, as can the contents of a Microsoft Outlook or other electronic address book. Collecting this type of information will become increasingly important for the U.S. Intelligence Community and the intelligence services of U.S. adversaries.

This new type of highly personalized influence operations, supported by computer network technology, emphasizes the importance of employing good OPSEC to protect sensitive but unclassified information. It also emphasizes the importance of having up-to-date anti-virus software and training computer users to delete emails from individuals they do not know. This is because many Trojan horse programs, which are often transmitted as attachments to emails, exploit the contents of address books to mail itself to everyone in the address book once the email attachment is executed. Malicious code has become increasingly sophisticated to the point that it is now possible to execute malicious code by simply opening an email message, without having to open an attachment.

It is difficult to derive lessons on computer network defense (CND) from OIF. There was a slight increase in cyber attacks against DoD systems during the initial days of OIF, but nothing significant was reported. It is not know whether these attacks were simple scanning or attempted network penetrations. The bottom line is there was "nothing systemic that could be tied back to the enemy."[39] The military should not take comfort in the fact that so far during OIF there have been no know penetrations of DoD systems attributable to the enemy. The Iraqis never possessed a sophisticated network attack capability. Richard Clarke, former Chairman of the President's Critical Infrastructure Board, recently issued a stern warning stating, "IT [information technology] has always been a major interest of al Qaeda. We know that from the laptops we have . . . that we've recovered that have hacking tools on them . . . it is a huge mistake to think that al Qaeda isn't technologically sophisticated, a fatal one. They are well trained; they are smart. They proved it on 9/11 with one style of attack, and they can prove it again."[40]

The Post-War Information Environment in Iraq

One of the harshest criticisms of the Bush administration's handling of the GWOT is the accusation there was no comprehensive plan to stabilize the situation upon the conclusion of major combat operations. To say there was no plan is too harsh; to say the plan was lacking is an understatement. This was particularly so in the area of influence operations, which dominate IO in stability operations. Conducting effective influence operations has proven difficult in the aftermath of war. To better understand the problems, it helps to look at the complex information environment in which the stability operations of OEF and OIF are taking place.

The 2004 presidential campaign in the United States had a major impact on shaping the global information environment. The American media has a tendency to focus on controversial political themes, which boosts sales and advertising but also creates false impressions impacting the perceptions of America's enemies. Democratic candidate John Kerry repeatedly accused President Bush of failing to form a legitimate coalition for OIF. Many Kerry critics claim this weakened Coalition cohesion and weakened political support in their home countries for America's Coalition partners in OIF. Conservatives charge that Kerry's Senate voting record showed he was weak on national defense and therefore would be weak against terrorism. Both the Democrats and Republicans politicized the Congressional 9/11 hearings. Private activists on both sides of the political aisle have joined the skirmish. Controversial film producer Michael Moore produced the movie *Fahrenheit 9-11*, which claimed President Bush fabricated the evidence for war against Iraq for his own political gain and for financial gain in the oil industry. Conservatives countered with a book titled *Unfit for Duty*. Containing wartime accounts of former navy swift boat veterans who served in Vietnam with John Kerry, the book claimed Kerry fabricated many details about his service in Vietnam to further his political career and accused him of aiding and abetting the enemy. For a terrorist studying the American political scene, the United States must appear like a country in total chaos. Retired General Tommy Franks, who commanded Coalition forces in the war against Iraq, summed up the negative impact of the political ruckus like this: "The terrorist read our papers and see our news, and the enemy is being given to believe they are winning."[41] Such a fray, whether intentional or not, hinders the public diplomacy, public affairs, and PSYOP efforts of the United States.

Some of the worst difficulties the DoD has encountered were self-inflicted. The U.S. military has a well-deserved reputation for kindness towards prisoners of war. This reputation received a black eye as a result of the Abu Ghraib prisoner abuse scandal. Photographs of American military guards abusing prisoners at the Abu Ghraib prison managed to make their way into the hands of the media. Not surprisingly, a media feeding frenzy ensued. The harm to influence operations caused by the Abu Ghraib prisoner abuse scandal is immeasurable. America's adversaries quickly seized the opportunity to use the photographs as a potent propaganda weapon. The pictures were widely distributed via Internet websites along with stories, some true and some fictitious, about the abuse of Iraqi prisoners by Americans. Al Qaeda has reportedly used the prisoner abuse scandal as a recruiting tool.

The graphic Abu Ghraib photos show nude and partially nude male Iraqi prisoners being abused in multiple fashions. Perhaps the most infamous photo is of army private first class (PFC) Lynndie England holding a leash attached to a dog collar on a nude Iraqi male prisoner. Another shows England smiling and pointing at the genitals of a nude male prisoner. What these photos show is not perhaps as significant as what they certainly imply—namely a total lack of understanding of (or a total disregard for) certain social sensitivities in Arab culture. Why was a female soldier allowed close contact with male prisoners, something not normally done in civilian prisons in the Western society? Why were the male prisoners nude before a female? Why the leash? Why, why, and why? Were the photos to be used to intimidate prisoners under interrogation in some ill-conceived scheme, as some of those charged with abuse claim, or do they portray the acts of demented individuals having some sort of perverted fun? At the time of this writing, one American soldier has already been convicted of prisoner abuse by a military court martial, another has confessed to crimes against the prisoners, and PFC England's trial is underway. Regardless of the outcomes of these and future trials, they will do little to assuage the damage done to America's ability to conduct effective influence operations in the GWOT. Effective influence operations, particularly PSYOP, rely heavily upon maintaining credibility. The visual impact of the abuse photos on the

Arab psyche, along with allegations that the Bush administration tried to suppress news of the abuse, have hurt America's credibility and will hinder future PSYOP, public affairs, and public diplomacy efforts.

Spreading the Good News

A recurring allegation by senior officials of the Bush administration has been that the media is focused on reporting the negative developments in Afghanistan and Iraq, while ignoring the positive developments. The same opinion was voiced by Virginia Congresswoman Jo Ann Davis. Davis said, "The overbearing negativity of the national press is not indicative of what is really going on in Iraq. The opposition we are facing is just a small minority (albeit it a deadly one) . . . the doom and gloom being painted by the national media is not only inaccurate at times, but it is detrimental to the morale of our troops."[42]

While the administration has complained about negative reporting, they have yet to develop an effective strategy for dealing with it. The embedded media, which covered the action so well during major combat operations, have all but disappeared from the television news. The DoD would do well to encourage the presence of embedded media during stability operations—admittedly a difficult proposition when one considers the media's first focus is on profit. Unfortunately, blood and destruction stories are more profitable than blissful scenes of reconstruction and progress. This fact has caused Coalition operations in Afghanistan to nearly disappear from the news, a sign that the security situation is greatly improved. Operations in Afghanistan are replete with success stories, yet these are buried in the news if covered at all.

Even greater untold success stories can be found in Iraq, particularly in the northern Kurdish region. While recovery seems to drag on in the rest of Iraq, the pace of recovery in the Kurdish north is astounding. As noted military commentator Ralph Peters notes about the north of Iraq, they have rule-of-law, financial efficiency, a prospering university system, and women can walk freely on the relatively safe streets. But as Peters points out, "If only the Kurds had a disaster or two, then someone might tell their story."[43] To gain support at home and abroad, the Bush administration must discover how to get good news from Afghanistan and Iraq to the American and foreign publics.

While the United States may have difficulties getting the word out, its enemies have enjoyed great success in their utilization of the media. Unlike previous generations of terrorists, whose political aims led them to choose politically significant targets, today's terrorists are more indiscriminate. Their tactics make reconstruction and stability operations particularly difficult. Innocent civilians and politically insignificant targets are routinely attacked. As Cheryl Benard explains, "The new generation of terrorist does not spare unarmed humanitarians. They do not leave clinics, schools, and other benign civilian projects untouched. They destroy them especially, because they want civilians to suffer and reconstruction to fail. Fear and backwardness are a kingdom they can rule; healthy, secure, and prosperous populations have no use for them."[44]

This new approach by the terrorists in Iraq has led to targeted kidnappings of innocent humanitarian workers and other civilians, in an effort to compel their governments to leave the U.S.-led Coalition and withdraw their forces from Iraq. Photographs and video of the brutal beheading of hostages, which have appeared in the international media and on numerous Islamist websites, have reinforced the terrorists' attempts to spread psychological terror. It was just such pressure that helped encourage Spain and the Philippines to withdraw their military forces from Iraq. The U.S. has yet to demonstrate any ability to counter this strategy through public affairs, public diplomacy, or PSYOP. Until it does, there is no reason to believe the terrorists will not continue to employ this approach to destroy Coalition cohesion.

How to convince the international media to refrain from excessive coverage of terrorist actions is a critical challenge for the Coalition. As Ollie Ferreira explains, "terrorism cannot succeed if it lacks media coverage, especially hostage situations, because without cooperative media coverage the terror act does not reach its intended audience."[45] Just convincing the American media to temper its coverage of terrorist actions would be a major accomplishment. Convincing the foreign media, particularly in the Arab world, is a more formidable challenge.

The Internet Dilemma

Al Qaeda very effectively employs the Internet to further its cause. According to the Christian Science Monitor, there are around four thousand Internet websites that support terror in one fashion or another.[46] Above all, al Qaeda and other Islamic terror organizations use the Internet to spread propaganda and psychological terror. "Since 11 September 2001, the organization has festooned its websites with a string of announcements of an impending 'large attack' on U.S. targets. These warnings have received considerable media coverage, which has helped to generate a widespread sense of dread and insecurity among audiences throughout the world and especially within the United States."[47] It seems today that it only requires a rumor of an attack to raise the national terrorism threat level in the United States. Besides being very costly, this is clear evidence of the powerful impact of psychological terror on American society today.

Al Qaeda and other terrorist organizations have used Internet for data mining open-source information to develop targets during attack planning; soliciting donations from supporters; recruiting and mobilizing sympathizers and new recruits; and collecting and transmitting information on bomb construction and weapons.[48] Command and control information is also passed via the Internet. Coded instructions may be placed on innocuous-appearing websites or embedded in emails, web pages, and photographs using sophisticated steganography techniques. To make matters more difficult, websites may be easily opened and closed or relocated to a new location on the web, making them nearly impossible to trace. Stopping such activities will prove difficult if not impossible. So it appears the only viable solution for the United States and its allies in this war is to seek new technologies that improve their capabilities to monitor these websites and to rapidly exploit the information gained.

Struggling to Win Hearts and Minds in the Muslim World

To Muslim parents, terrorists like bin Laden have nothing to offer
their children but visions of violence and death. America and its
friends have a crucial advantage—we can offer these parents a
vision that might give their children a better future.

9/11 COMMISSION REPORT[49]

Well before publication of the 9/11 Commission report, the USG had already embarked upon several public diplomacy projects to send a message of hope to parents and others in the Muslim world. These have not been as successful as had been hoped. One of projects is Alhurra, a new Arabic-language satellite television news channel. Alhurra began broadcasting to millions of viewers in twenty-two countries across the Middle East on 14 February 2004.[50] It reports mainly news and information of interest to the Muslim world. Alhurra (Arabic for "the free one") is funded by the USG, and run under the supervision of the Broadcasting Board of Governors (BBG). The BBG is a federal agency that serves as a sort of buffer between the USG and the broadcasters, in order to protect the integrity of the broadcasts from undue USG influence. Critics charge the BBG has not fully succeeded in this role.

Alhurra is available to Middle East viewers via TV satellite broadcasts on Arabsat and Nilesat, which together cover nearly the entire region. It has enjoyed some success in Iraq, where a recent survey indicates that 61 percent of Iraqi adults watched the network during the week prior and that 64 percent found the news to be very or somewhat reliable.[51] In April 2004 Alhurra added a second channel focused specifically on Iraq. The *Jordan Times* complained, "While some expected the new station to be an important addition to the plurality of opinions available to the Arab public, its first day of broadcasting confirmed what the skeptics have been saying all along: What the U.S. needs to do is change its policy, not its media strategy."[52] The complaint focused on the first person to be interviewed on the new network, none other than President George W. Bush. Critics complained that Bush used worn-out political rhetoric and his interviewer failed to address critical questions about Israel. Al Sawa, a post-11 September Arabic-language radio station also managed by the BBG, has received similar criticism. It is evident that the USG's actions towards Israel speak louder than its words to the Muslim world. How to overcome this problem remains a mystery that must be solved if the United States hopes to successfully implement the recommendations of the 9/11 Commission and to gain credibility with its target audiences.

Steven A. Cook of the Council on Foreign Relations suggests now that the transfer of political authority to the interim government in Iraq is complete, it would be a good time to change the content of Alhurra to focus on how the democratic process functions. Cook suggests that Alhurra should be a sort of C-Span for the Muslim world, focusing on the U.S. Congress, which would help counter many of the " myths and conspiracy theories about how American policy is developed and articulated."[53] Cook goes on to suggest that without change Alhurra cannot hope compete with Al Jazeera and Al Arabiya television, its main market competitors in the Muslim world.

Another USG program aims to reach young Arabs. Published by the State Department, *Hi* magazine targets Arabs aged eighteen to thirty-five with articles about social and cultural issues in America.[54] The magazine debuted in July 2003 with an initial circulation of about fifty thousand and an eventual goal of two hundred and fifty thousand. A recurring complaint from Arabs has been that the magazine focuses on American life and culture in attempt to improve its tarnished image, but ignores most important issue of all, which is the U.S. "double-standard foreign policy," a term referring to the perceived favoritism shown by the United States in the Arab-Israeli dispute.[55] It is too early to determine whether or not *Hi* magazine will accomplish its aim. We will have to wait until the young generation of readers begins to assume leadership roles in their respective countries. However, it at least represents an attempt by the USG at a long-term approach to winning hearts and minds in the Arab world.

The Iraqi Media Network (IMN), which has now been renamed the Al Iraqiya network, is another USG program whose results have been questionable. The IMN's mission was to establish the first independent radio and television network in post-Saddam Iraq. The program began through a DoD contract with Scientific Applications International Corporation, funded by the office of the Assistant Secretary of Defense for Special Operations and Low Intensity Conflict, the same office responsible for DoD psychological operations.[56] The sponsoring office's PSYOP connection had skeptics questioning the independence of the IMN and the credibility of its content. Criticism increased when Coalition Provisional Authority (CPA) leader Paul Bremer placed controls on IMN content. In June 2004, in response to growing violence in Iraq, Bremer issued the Iraqi independent media, including IMN, "a nine-point list of 'prohibited activity' that included incitement to violence, support for the Baath Party, and publishing material that is patently false and calculated to promote opposition to the occupying authority."[57] This and other perceived interference by Bremer led to the resignation from IMN of the highly experienced producer Don North, who went to Iraq to help form IMN. North charged the IMN, "just became a mouthpiece for the Coalition and the Iraqis didn't find it credible."[58] North delivered a long list of charges against the CPA in a December 2003 article titled, "One Newsman's Take on How Things Went Wrong." He claims the IMN had a "revolving door" of officials with no media experience; a lack of operating capital; was forced to air programming it objected to; was forced to cover endless CPA meetings; and was provided inadequate journalistic training which was actually turned over to Al Arabiya and Al Jazeera.[59]

While there are no easy solutions to the U.S. problem in communicating with the Muslim world, it is clear that establishing credibility must be a top objective. The United States is engaged in a hostile environment where actions speak louder than words. Situations like the Abu Ghraib prison scandal, the CPA's suppression of the fledgling independent media in Iraq, the U.S. Army's stand-off with hostile Shiite militia at the Imam Ali mosque in Najaf, one of the most holy Shiite shrines, and prisoners held without charges in Guantanamo Bay make any hope of near-term public diplomacy success for the USG appear remote. If the USG truly desires to introduce democracy to the Middle East, the independent media in Iraq and Afghanistan must be afforded at least a degree of the freedom of the press afforded the media in the United States. How to handle unrestricted media while simultaneously engaging in military conflict in an area is a difficult problem. The hand over of power from the CPA to the Iraqi interim government makes this particularly challenging in Iraq, where at least in theory the problem now belongs to the Iraqi government. A long-term strategy of how to reach the Muslim world through public diplomacy and public affairs is an essential element for success in the GWOT.

IO Capabilities and Related Activities in the Post-War Period

During stability operations, the IO capabilities of military deception and electronic warfare have extremely limited roles and are therefore not discussed here. Likewise, due to the extremely limited open source information available pertaining to classified DoD computer network operations, they are not discussed in this section

OPSEC

After receiving much scrutiny from Secretary of Defense Rumsfeld after the 11 September attacks and during major combat operations in OEF and OIF, DoD OPSEC appears to have improved significantly. One example of OPSEC success was President Bush's surprise visit to Iraq on Thanksgiving Day in 2003. The surprise was complete, although the trip received some criticism from the media because the flight information was apparently falsified for security reasons.[60] Global Security, a military watchdog organization, follows U.S. global military operations. Among other things, the organization tracks the deployments of U.S. military units and posts this information on their website. Global Security recently reported, "The events of 11 September 2001, and the Global War on Terrorism have made such an effort [tracking the movement of U.S. military units] significantly more difficult as the military seeks to improve operational security and to deceive enemies and the media as to the extent of American operations."[61]

One result of the renewed DOD focus on OPSEC is the creation of the Joint OPSEC Support Element (JOSE), a subordinate element of the Joint Information Operations Center (JIOC) in San Antonio, Texas. The JOSE was organized early in 2004 and showcased at the National OPSEC Conference in June 2004. "A small team of OPSEC professionals drawing on myriad operational experiences, the JOSE provides OPSEC support to DoD activities in the areas of Program Development and Training, Survey Support, and Plans and Exercise support."[62] Creating the JOSE gives the DoD, and particularly the Combatant Commanders, access to OPSEC support similar to that provided by the Interagency OPSEC Support Staff (IOSS), which has supported the U.S. National OPSEC Program for a number of years.

Combined with the IO planning support already provided by the JIOC, the JOSE represents a significant capabilities enhancement and a much-needed capability in the GWOT.

PSYOP

The DoD is attempting to improve its PSYOP capabilities based upon the lessons learned during OEF and OIF. The commander of U.S. Special Operations Command (USSOCOM) is the Combatant Commander responsible for PSYOP. Troubled for years by a paucity of active duty PSYOP forces, the command is finally beginning to expand both the active and reserve component PSYOP forces. Testifying before the Senate Armed Services Committee on his nomination for appointment to commander of USSOCOM, Lt. Gen. Bryan D. Brown stated, "Due to the high demand for PSYOP forces, USSOCOM is in the process of growing its PSYOP force structure by adding two active duty regional companies and four reserve component tactical companies. This year the command also proposed an advanced technologies concept demonstration aimed at improving PSYOP planning tools and long range dissemination into denied hostile areas. In addition, USSOCOM is creating a seventy person Joint PSYOP Support Element (JPSE), to provide dedicated joint PSYOP planning and expertise to geographic combatant commanders, strategic command, and the Secretary of Defense."[63]

One new PSYOP capability sought appears is global shortwave broadcast. This indicates the DoD's recognition that many of the countries where the GWOT is being or will be fought have limited communications infrastructure, meaning the populations in these countries most often rely upon radio for their primary access to the media. In April 2004, the JPSE solicited for contractors to build a new PSYOP support capability. The solicitation stated, "The requirement is to conduct short-wave broadcasting transmission of informational news for coverage over wide area via shortwave for the Joint Psychological Operations Support Element for the United States Special Operations Command."[64] This and much more PSYOP innovation is needed for the GWOT.

Civil Affairs

There is no questioning the tremendous impact of civil affairs operations, particularly during stability operations, when "winning hearts and minds" is so important. Nowhere has the challenge been more difficult than in Afghanistan, where the greatest civil affairs effort since World War II is underway.

As early as the one-year anniversary of the beginning of OEF, the Coalition had improved Afghanistan's infrastructure to levels never imagined before the war. The statistics were impressive:

Ten water projects were completed during the first six months of 2002. These included eighty-three wells, benefiting approximately 260,000 Afghans. An additional sixteen new water projects had been approved. Demining teams from Norway, Britain, Poland, and Jordan helped clear mines from more than 1.8 million square meters of terrain. U.S. Army Civil Affairs troops completed sixty-one school repair projects with plans for forty-four more. The U.S. provided ten million textbooks and four thousand teacher-training kits. Canada, Greece, Belgium, and Iceland delivered sixty metric tons of goods. Jordan built a hospital in Mazar-e-Sharif that treated more 105,000 patients. Belgium led a multinational humanitarian assistance mission that delivered ninety metric tons of UNIMIX to starving children in Afghanistan. The U.S. jointly funded the measles vaccinations of more than four million children."[65]

The size and cultural make-up of Afghanistan has forced civil affairs forces to innovate. Instead of the traditional force structure where there is one main civil affairs headquarters for a large command, like a Joint Task Force, the civil affairs force structure in Afghanistan has been organized into "provisional reconstruction teams. (PRTs), [which] are teams of international civilian and military personnel working in Afghanistan's provinces to extend authority of the Afghan central government and to facilitate development and reconstruction."[66] The military portion of a PRT consists of civil affairs, PSYOP, special operations, and regular army forces.[67]

The first PRT in Afghanistan was formed on 31 December 2002. As of April 2004, there were twelve, of which nine were U.S.-led and three were led by Coalition partners.[68] The idea behind the PRT concept is to extend security and reconstruction beyond the Afghan capital Kabul's relatively secure surroundings out into the remote, mountainous countryside where tribal culture dominates. According to Lt. Col. John Lineweaver, who commanded a PRT headquartered in Heart (Afghanistan), the operational area of a PRT resembles a hub and spokes with the PRT headquarters in the center and a series of safe houses extending the capabilities outward.[69]

Civil affairs efforts have had tremendous success in Afghanistan. According to Lt. Gen. John R. Vines, Coalition Forces Commander in Afghanistan from September 2002 to October 2003, because so many of Afghanistan's people live in relative isolation, "the central government had never had any influence on the daily lives of the majority of the people. So when a provisional reconstruction team provides electricity to a village, or in some cases farming equipment, it has a huge impact on their daily lives."[70]

The success of the overall reconstruction efforts in Afghanistan is highly dependent upon extending security throughout the country. The efforts appear to be paying off. Voter turnout for the presidential election was high; the national road system is well on its way to recovery after years of neglect; and Afghanistan has signed a grant and credit agreement with the World Bank.

The civil affairs operations in Iraq differ significantly from those in Afghanistan. Unlike Afghanistan, elected officials in Iraq must often informally share power with local tribal leaders. This makes implementing civil affairs in certain areas more difficult, because it requires negotiations with the elected officials and the tribal leaders. Iraq has many more urbanized areas and a more highly developed infrastructure to rebuild and maintain than Afghanistan. Restoring electricity-generation capacity has been a difficult challenge and is a particular concern for Iraq's city dwellers. So has unemployment. Radical groups have attacked Iraq's infrastructure, particularly oil pipelines, in an attempt to slow recovery and destabilize the area. The often unstable security situation has slowed down big reconstruction projects, which exacerbates the unemployment situation among Iraqi citizens. The attacks and intimidation against relief and reconstruction workers have been particularly costly in terms of slowing recovery.

In Iraq, more Coalition military forces are interspersed with civilians than in Afghanistan. For this reason, "the Army has assigned a civil affairs team to almost every battalion to take charge of reconstruction projects and set up neighborhood councils to get Iraq back to normal."[71] Because of the security situation, many of the reconstruction projects in Iraq have been small, local ones that are often performed directly by military units. Ambassador Paul Bremer told Fox News on 4 July 2004, that eighteen thousand reconstruction projects had been completed in Iraq in the past year.[72]

When comparing the levels of success of civil affairs operations in Afghanistan and Iraq, several things become readily evident. First, the unstable security situation in Iraq is slowing recovery progress. The attacks against critical economic infrastructure, aid workers, and other humanitarian workers are a particularly vexing problem. Additionally, the departure of Coalition partners like Spain and the Philippines in wake of terrorist attacks and intimidation hampers security efforts, further hindering recovery. Establishing and maintaining security is essential for effective civil affairs operations. Additionally, the old problem of failing to understand cultural and regional differences in an area of operations continues to hamper DoD IO.

Public Affairs

In the war of words, the DoD has come up short in the GWOT. This has been particularly so in Iraq, where the Bush administration had much less public support for the war than in Afghanistan. One problem for the DoD has been media relations. Secretary of Defense Rumsfeld has displayed a tendency to be combative and rather flippant with the media when they question his policies or thinking. Many in the media have, in turn, assumed a combative posture towards Rumsfeld. This cannot but hamper DoD public affairs efforts.

One has to wonder why, with so many war opponents charging that the administration failed to form a legitimate Coalition in Iraq, global audiences rarely saw a non-American Coalition spokesperson addressing the media from the war zone. Instead the monotone, straight-faced, talking head, U.S. Army Brig. Gen. Mark Kimmitt, was the usual spokesman appearing before the international media. Compare this to the NATO war in Kosovo, where multiple spokespersons from different Alliance members standing before the cameras were common. The Coalition may have benefited from a similar approach in Iraq.

The DoD apparently forgot a lesson learned early in the GWOT. As discussed previously, during the major combat operations in OEF and OIF, DoD officials held several press briefings specifically aimed at exposing enemy D&D methodology and explaining in detail how they could prove that the purported evidence was faked. Most of the evidence came from comparing overhead imagery of the target areas before and after the attacks. This approach was fairly effective in neutralizing some of the enemy propaganda that might have otherwise required much greater effort to overcome. Yet during stability operations, public affairs operations were never employed in a similar manner to discredit the enemy. There were numerous opportunities to do this, but none better than in the case of militants attacking Coalition forces from inside mosques in the towns of Najaf and Fallujah. While the U.S. Army suffered accusations of Geneva Convention violations at the Abu Ghraib and Guantanamo Bay prisons, many blatant violations by the enemy went nearly unnoticed by the international media. Incriminating evidence could have easily been gathered by combat camera teams filming at the mosques and other locations where violations occurred. If, in fact, such video was recorded, then the DoD must attempt to ascertain why the media never aired it.

On a positive note, the DoD has made some progress in solving a public affairs problem it has struggled with for years—quickly transmitting information internally and making it available to the media in a timely fashion. This ability is particularly important for information that may not be favorably received by the public, such as news on the accidental wounding or killing of civilians. It is important for the public affairs forces to break bad news to the media before the enemy seizes the opportunity to exploit it for propaganda purposes. A new system developed for combat camera teams helps do just this. In the past, combat camera video was filmed and then transported by ground or air transportation to higher military headquarters for further processing. With the newly fielded Joint Camera Imagery Transmission Satellite System (JCCITSS), video is now relayed by satellite in near-real time to higher headquarters and ultimately to the Pentagon.[73] The JCCITSS system is an indication that DoD public affairs technology may now be moving towards a position on par with the technology used by civilian media, an area where the DoD has lagged behind for years. It is important that the United States continue to take advantage of technological superiority to enhance the influence operations capabilities that will be so critical to success in the GWOT.

Conclusions from IO Lessons Learned in OEF and OIF

To maintain information dominance, we must commit to improving our ability to influence target audiences and manipulate our adversary's information environment. Continued development of these capabilities is essential.

GENERAL TOMMY FRANKS[74]

One of the easiest conclusions to draw from the employment of IO in the GWOT is that the ability to effectively command and control (C2) seamless IO, particularly at the strategic and operational levels, remains elusive. This is not for lack of effort, but simply reflects how difficult it is to plan and synchronize the myriad tasks that must be successfully performed in order to accomplish the IO objectives in major military operations. J. Michael Waller provided a wonderful summary of the challenge.

> Many of the U.S. public-diplomacy and information operations in support of the war effort have been piecemeal, tactical, and mostly reactive instead of strategic, comprehensive, and anticipatory. A long-term strategy has yet to be developed, according to administration officials. That, critics say, leaves the enemy to define the terms of debate and severely complicates U.S. diplomacy and military planning.[75]

Unfortunately, we've heard similar criticism of IO since the publication the DoD Joint IO Doctrine, Joint Publication 3-13, in October 1998. Developing a comprehensive national IO strategy for the GWOT should remain a top priority for the United States.

Coordinating DoD IO is particularly difficult when it requires the cooperation and synchronization of activities between organizations in the interagency, which is usually the case. One example of this occurs far too frequently—the seeming inability of the White House, Department of State, and DoD to harmonize the themes and messages in their media releases. Each has a robust media capability, yet each appears unable to effectively coordinate with the other. The frequent result is a collection of mixed messages that confuse friends and adversaries alike as to the plans and intentions of the United States. This is a major obstacle to convincing others to act in a manner that supports U.S. national security objectives. It may be the greatest IO-related problem the U.S. needs to solve in the GWOT. The media can be a strategic enabler in a number of ways: to communicate the objective and end-state to a global audience, to execute effective psychological operations (PSYOPS), to play a major role in deception of the enemy, and to supplement intelligence collection efforts.[76]

The failure of the Pentagon to successfully establish the Office of Strategic Influence (OSI) is another example of coordination difficulties at the highest levels in the USG. The situation was further complicated because it was rumored that one of the office's missions was to engage in so-called black propaganda operations. This is the term used to describe the surreptitious placing of false information in the foreign media in order to influence the behavior of adversaries. Part of the controversy over using black propaganda is misinformation planted abroad can reach and influence American audiences via today's global communications. This goes against the grain of American culture and raises a number of legal issues as well. As J. Michael Waller observed,

> [T]he Department of Defense set up a new Office of Strategic Influence (OSI) to run information operations abroad in support of U.S. strategic-

defense goals. However, OSI hardly had gotten off the ground when the bureaucracy and turf-jealous senior officials leaked misleading, inflammatory, and utterly dishonest stories that falsely portrayed the OSI as intending to plant 'disinformation' in the [American] press.[77]

The OSI fiasco was a blow to the United States' ability to conduct strategic PSYOP and to the credibility of U.S. strategic communications throughout the world. Of the OSI, Secretary of Defense Rumsfeld said after it was disassembled,

> And then there was the office of strategic influence. You may recall that. And "oh my goodness gracious isn't that terrible, Henny Penny the sky is going to fall." I went down that next day and said fine, if you want to savage this thing fine I'll give you the corpse. There's the name. You can have the name, but I'm gonna keep doing every single thing that needs to be done and I have.[78]

Functions of the defunct OSI have ostensibly been assumed by the White House Office of Global Communications (OGC), which lists countering propaganda and disinformation as part of its mission. Whether or not the new office will be effective has yet to be determined. However, the reluctance of the USG to establish a strategic PSYOP organization does not bode well for the GWOT.

The Chairman of the Joint Chiefs of Staff, Gen. Richard B. Myers, summed up IO in Afghanistan: "It took too much time to put together the team. We missed the opportunity to send the right message, sometimes we sent mixed signals, and we missed opportunities as well."[79] Obviously, coordination efforts, particularly with other government organizations in the so-called interagency continue to hamper DoD IO. The C2 of IO during operations in Iraq appeared to be better coordinated. Apparently the IO lessons from OEF were applied with some success in OIF. This is particularly apparent when one compares the restrictive media policy applied by the DoD in Afghanistan with the embedded media employed during major combat operations in Iraq.

The DoD has recently taken steps improve the C2 of IO. A 2003 change to the Unified Command Plan, the document that describes the roles and missions of the U.S. military's major combatant commanders, created a Joint Force Headquarters for IO (JFHQ-IO) under the command of the U.S. Strategic Command (USSTRATCOM). This change effectively consolidates the responsibility for DoD IO under one Combatant Commander. The effectiveness of the change has yet to be demonstrated, but in theory it should improve coordination of IO, at least within the DoD.

Another clear lesson is that the USG has yet to discern how to effectively deal with the profound psychological impact of terrorism and the attacks on 11 September. The attacks "had deep human, economic, and psychological impacts. The terrorists were not deterred by our overwhelming military superiority, in fact, for that day at least, they made it irrelevant. Traditional concepts of security, threat, deterrence, warning, and military superiority don't completely apply against this new strategic adversary."[80] According to the Chairman of the Joint Chiefs of Staff, "This is a new kind of war. The military may not be decisive."[81] These observations emphasize the criticality of improving coordination in the interagency environment and enhancing the DoD's IO capabilities.

The American medical community has observed an increase in anxiety over terrorism. As Todd Zwillich observes,

> In their zeal for physical security, U.S. policy makers may be ignoring terrorists' most potent weapon: fear. Several experts say that the government has not done enough to teach Americans how to reduce or manage their fear of an attack, making it more likely that the next strike will cause the mayhem and disarray that terrorists crave.[82]

The media's craving for video footage depicting acts of terrorism has exacerbated the problem. The constant bombardment of hostage executions and improvised explosive device (IED) attacks seen on television cannot but help the terrorist cause, whether intentionally or unintentionally. One of the most blatant examples is a CNN report showing a series of videos of IED explosions that were provided by Islamic radicals to Michael Ware, an Australian journalist with *Time* magazine. This particular video series clearly indicates how the terrorists in Iraq have refined a methodology for producing live video of terrorist attacks.[83] It is also sends a potent psychological message to the public in the United States and its Coalition partners.

Americans living in the post–11 September era have a changed worldview— a change that occurred literally overnight. The false sense of security provided by America's relative geographic isolation for over two hundred years is forever gone. Americans today live in a new society. In the wake of 11 September, the U.S. enjoyed tremendous international support for its military operations in Afghanistan. However, the Bush administration's new policy of preemptive engagement to strike potential enemies before they strike the U.S., as was used to help justify the invasion of Iraq, has eroded that support. The strategic communications efforts of the U.S. have failed to garner international support for stability operations in Iraq. The Patriot Act was enacted to help law enforcement agencies root out terrorists operating in the United States, but many citizens believe the act infringes upon their constitutional rights. Increased security at airports, public buildings, and public events has changed the daily routines of Americans. An anthrax scare nearly crippled the U.S. postal system and slowed government operations when contaminated letters were sent to members of Congress. A pair of deranged snipers created massive public fear and chaos in the greater Washington, D.C. area over the course of a month in September and October 2002. A major electric power failure in the northeast U.S. raised immediate fears of a terrorist attack. The dread of terrorism is now firmly fixed in the American psyche, so the terrorists have scored a major victory. How should America respond?

A part of the response must be the continuation of diplomatic, military, legal, intelligence, and technical efforts to cut off terrorist funding and logistics throughout the world. It is also essential that the USG develops a comprehensive national IO strategy for the GWOT and continues efforts to develop an effective strategic communications program that clearly explains U.S. actions and clearly conveys America's intentions to its friends and enemies alike. While the USG continues to struggle with a strategy for winning the GWOT, military stability operations will continue in Afghanistan and Iraq with no foreseeable end. President Bush has declared Iraq as the central stage for the GWOT, due to the large number of terrorists that have entered the country to engage Coalition forces. It is essential for the DoD to quickly and thoroughly analyze the IO lessons learned and develop solutions to known problems. DoD IO for the remainder of OEF and OIF will focus mainly on influence operations. It would be wise to focus on quickly analyzing influence operations lessons learned to ensure that mistakes in this critical area of IO are not repeated. The success of public diplomacy, public affairs, and DoD influence operations will play key roles in determining the futures of Afghanistan and Iraq.

CHAPTER 5

Cyberterrorism: Hype and Reality

Dr. Maura Conway
Lecturer, Dublin City University

The term "cyberterrorism" unites two significant modern fears: fear of technology and fear of terrorism. Both of these fears are evidenced in the following quote from Walter Laqueur, one of the most well known figures in terrorism studies: "The electronic age has now made cyberterrorism possible. A onetime mainstay of science fiction, the doomsday machine, looms as a real danger. The conjunction of technology and terrorism make for an uncertain and frightening future."[1] It is not only academics that are given to sensationalism. Cyberterrorism first became the focus of sustained analysis by the U.S. government in the mid-1990s. In 1996 John Deutch, former director of the Central Intelligence Agency (CIA), testified before the Permanent Subcommittee on Investigations of the United States' Senate Governmental Affairs Committee:

> International terrorist groups clearly have the capability to attack the information infrastructure of the United States, even if they use relatively simple means. Since the possibilities for attacks are not difficult to imagine, I am concerned about the potential for such attacks in the future. The methods used could range from such traditional terrorist methods as a vehicle-delivered bomb—directed in this instance against, say, a telephone switching center or other communications node—to electronic means of attack. The latter methods could rely on paid hackers. The ability to launch an attack, however, are [*sic*] likely to be within the capabilities of a number of terrorist groups, which themselves have increasingly used the Internet and other modern means for their own communications.[2]

Both the popularity and, to some extent, the credibility of such scenarios was given a boost by the entertainment industry. Hollywood, eager to capitalize on the cyberterrorist threat, released the James Bond film *Goldeneye* in 1995, which highlighted the threat. Other sectors were quick to follow with the publishing industry introducing Tom Clancy and Steve R. Pieczenik's *Net Force* series in 1998. As Ralf Bendrath has pointed out:

> Sometimes it is hard to tell what is science and what is fiction. Winn Schwartau, for example, the rock manager turned preacher of "information warfare" who runs the famous website infowar.com, has testified several times as an IT security expert before Congress, and has written two novels on cyber-terror. Even renowned cyber-war theoreticians like John Arquilla have not hesitated to publish thrilling cyber-terror scenarios for the general audience. But these works are not only made for entertainment. They produce certain visions of the future and of the threats and risks looming there.[3]

In 1998 the Global Organized Crime Project of the Center for Strategic and International Studies in Washington, D.C. published a report titled "Cybercrime, Cyberterrorism, Cyberwarfare: Averting an Electronic Waterloo." This was the first major academic contribution to the field. The document's authors view cyberterrorism as a subspecies of information warfare (IW). This is because information warfare is a form of asymmetric warfare and is therefore viewed as an eminently suitable terrorist strategy. Cyberterrorism has since come to be viewed as a component allied to offensive information warfare, but one that has a direct corollary in traditional, physical, non-information based warfare (i.e., classical political terrorism). In other words, cyberterrorism is recognized as having links with traditional terrorist tactics, but may be viewed as a new strategy employing new tools and exploiting new dependencies.

Although the authors of the CSIS report fail to provide a definition of what it is they mean by "cyberterrorism," they are at pains to illustrate its potentially disastrous consequences:

> A smoking keyboard does not convey the same drama as a smoking gun, but it has already proved just as destructive. Armed with the tools of cyberwarfare, substate or nonstate or even individual actors are now powerful enough to destabilize and eventually destroy targeted states and societies. . . . Information warfare specialists at the Pentagon estimate that a properly prepared and well-coordinated attack by fewer than 30 computer virtuosos strategically located around the world, with a budget of less than $10 million, could bring the United States to its knees. Such a strategic attack, mounted by a cyberterrorist group, either substate or nonstate actors, would shut down everything from electric power grids to air traffic control centers.[4]

A focus on such "shut-down-the-power-grid" scenarios is increasingly a feature of analyses of the cyberterrorist threat.[5]

This chapter is thus concerned with explicating the origins and development of the concept of cyberterrorism with a view to separating the hype surrounding the issue from the more prosaic reality. This is more difficult than it may at first appear, however. Ralf Bendrath has identified three major stumbling blocks.[6] First, this debate is not simply about predicting the future, but is also about how to prepare for it in the present. The problem is that those involved in the debate cannot draw on either history or experience to bolster their positions, as a major cyberterrorist incident has never yet occurred. For this reason different scenarios or stories about the possible course of future events are providing the grounds on which decisions must be made. The upshot of this is that the multiple actors (e.g., government and opposition, the computer security industry, the media-entertainment complex, scholars, and others) with their various, and often times divergent, interests are competing with each other by means of their versions of the future, which are particularly subject to political exploitation and instrumentation.

A second, and related, problem is the nature of the space in which a cyberterrorist attack would occur:

> In the physical landscape of the real world, any action has its constraints in the laws of nature. . . . Cyberspace, in contrast, is a landscape where every action is possible only because the technical systems provide an artificial environment that is built to allow it. The means of attack therefore change from system to system, from network to network. This makes threat estimation and attack recognition much more difficult tasks.[7]

Bendrath's final point relates to the highly technical nature of the new threat and the constraints placed on social scientists and their ability to estimate the magnitude of that threat. Bendrath's solution is for social scientists to draw conclusions by looking at how the threat is perceived. "The way a problem is framed normally determines or at least limits the possible solutions for it."[8]

With this in mind, this paper seeks to excavate the story of the concept of cyberterrorism through an analysis of both popular media renditions of the term and scholarly attempts to define its borders. It must be stated at the outset that, in both media and academic realms, confusion abounds. This is startling, particularly given that since the events of 9/11, the question on everybody's lips appears to be "Is cyberterrorism next?".[9] In academic circles the answer is generally "not yet." The media are less circumspect, however, and policy makers appear increasingly to be seduced by the media's version of events. It seems to me that both question and answer(s) are hampered by the lack of certainty surrounding the central term. Let me begin by putting forward some concrete illustrations of this definitional void culled from newspaper accounts.

Cyberterrorists Abound

In June 2001 a headline in the *Boston Herald* read "Cyberterrorist Must Serve Year in Jail."[10] The story continued: "Despite a Missouri cyberterrorist's plea for leniency, a Middlesex Superior Court judge yesterday told the wheelchair-bound man 'you must be punished for what you've done' to Massachusetts schoolchildren and ordered him to serve a year in jail." The defendant pleaded guilty to "launching a campaign of terror via the Internet" from his Missouri home, including directing middle school students to child pornography websites he posted, telephoning threats to the school and to the homes of some children, and posting a picture of the school's principal with bullet holes in his head and chest on the Internet.

In December 2001 a headline in the *Bristol Herald Courier*, Wise County, Virginia read "Wise County Circuit Court's Webcam 'Cracked' by Cyberterrorists."[11] The webcam, which allows surfers to log on and watch the Wise County Circuit Courts in action, was taken offline for two weeks for repairs. "(Expletive deleted) the United States Government" was posted on a web page. However, the defaced page could only be seen by the Court's IT contractors; Internet surfers who logged on could only see a blank screen. The attack is thought to have originated in Pakistan or Egypt, according to the report. "This is the first cyberterrorism on the court's Internet technology, and it clearly demonstrates the need for constant vigilance," said Court Clerk Jack Kennedy. "The damage in this case amounted to a $400 hard drive relating to the Internet video server. The crack attack has now resulted in better software and enhanced security to avoid a [*sic*] further cyberterrorism." According to Kennedy, cracking (i.e., hacking for criminal purposes) can escalate to terrorism when a person cracks into a government- or military-maintained website; he said cyberterrorism has increased across the United States since the events of 9/11 and law enforcement has traced many of the attacks to Pakistan and Egypt. It was predicted that an escalation in attacks by hackers would occur in the aftermath of 9/11.[12] However, the predicted escalation did not materialize. In the weeks following the attacks, website defacements were well publicized, but the overall number and sophistication of these remained rather low. One possible reason for the non-escalation of attacks could be that many hackers—particularly those located in the U.S.—were wary of being associated with the events of 11 September and curbed their activities as a result.

In March 2002, Linkline Communications, described as "a small, but determined Internet service provider" located in Mira Loma, California received telephone and e-mail threats from an unnamed individual who claimed to have accessed—or to be able to access—the credit card numbers of Linkline's customers. He said that he would sell the information and notify Linkline's customers if $50,000 wasn't transferred to a bank account number that he supplied. The ISP refused to concede to the hacker's demands. "We're not going to let our customers, or our reputation, be the victims of cyberterrorism," said one of the company's founders. Linkline contacted the authorities and learned that the hacker and his accomplices may have extorted as much as $4 billion from other companies. The account was subsequently traced through Russia to Yemen.[13]

A similar incident had taken place in November 2000. An attack, originating in Pakistan, was carried out against the American Israel Public Affairs Committee, a lobbying group. The group's website was defaced with anti-Israeli commentary. The attacker also stole some thirty-five hundred e-mail addresses and seven hundred credit card numbers, sent anti-Israeli diatribes to the addresses, and published the credit card data on the Internet. The Pakistani hacker who took credit for the hack, the self-styled Dr. Nuker, said he was a founder of the Pakistani Hackerz Club, the aim of which was to "hack for the injustice going around the globe, especially with [*sic*] Muslims."[14] In May 2001 "cyberterrorism" reared its head once again when supporters of the terrorist group Laskar Jihad (Holy War Warriors) hacked into the website of Australia's Indonesian embassy and the Indonesian national police in Jakarta to protest against the arrest of their leader. The hackers intercepted users logging on to the websites and redirected them to a site containing a warning to the Indonesian police to release Ja'far Umar Thalib, the group's leader. Thalib was arrested in connection with inciting hatred against a religious group and ordering the murder of one of his followers. According to police, the hackers, the self-styled Indonesian Muslim Hackers Movement, did not affect police operations. The Australian embassy said the hackers did not sabotage its website, but only directed users to the other site.

It is clear that the pejorative connotations of the terms "terrorism" and "terrorist" have resulted in some unlikely acts of computer abuse being labeled "cyberterrorism." According to the above, sending pornographic e-mails to minors, posting offensive content on the Internet, defacing websites, using a computer to cause $400 worth of damage, stealing credit card information, posting credit card numbers on the Internet, and clandestinely redirecting Internet traffic from one site to another all constitute instances of cyberterrorism. And yet none of it could be described as terrorism—some of it not even criminal—had it taken place without the aid of computers. Admittedly, terrorism is a notoriously difficult activity to define; however, the addition of computers to plain old criminality terrorism is not.

The Origins of Cyberterrorism

Barry Collin, a senior research fellow at the Institute for Security and Intelligence in California, coined the term "cyberterrorism" in the mid-1980s.[15] The idea of terrorists utilizing communications technologies to target critical infrastructure was first noted more than two decades ago, however. In 1977, Robert Kupperman, then Chief Scientist of the U.S. Arms Control and Disarmament Agency, stated:

> Commercial aircraft, natural gas pipelines, the electric power grid, offshore oil rigs, and computers storing government and corporate records are examples of sabotage-prone targets whose destruction would have derivative effects of far higher intensity than their primary losses would suggest. Thirty years ago terrorists could not have obtained extraordinary leverage. Today, however, the foci of communications, production, and distribution are relatively small in number and highly vulnerable.[16]

Such fears crystallized with the advent of the Internet. The opening chapter of *Computers at Risk*, one of the foundation books in the U.S. computer security field, which was commissioned and published by the U.S. National Academy of Sciences in 1991, begins as follows:

> We are at risk. America depends on computers. They control power delivery, communications, aviation, and financial services. They are used to store vital information, from medical records to business plans to criminal records. Although we trust them, they are vulnerable—to the effects of poor design and insufficient quality control, to accident, and perhaps most alarmingly, to deliberate attack. The modern thief can steal more with a computer than with a gun. Tomorrow's terrorist may be able to do more damage with a keyboard than with a bomb.[17]

Nevertheless, cyberterrorism only became the object of sustained academic analysis and media attention in the mid-1990s. It was the advent, and then the increasing spread, of the World Wide Web (WWW) along with the vocal protestations of John Deutch, then Director of the Central Intelligence Agency (CIA), as to the potentiality of the Web as a terrorist tool and/or target that kick-started research into the phenomenon of cyberterrorism in the United States.

From Real World Terrorism to Cyberterrorism

It has been pointed out that if you ask ten people what "cyberterrorism" is, you will get at least nine different answers.[18] This discrepancy bears more than a grain of truth, as there are a number of stumbling blocks to constructing a clear and concise definition of cyberterrorism. Chief among these are the following:

- A majority of the discussion of cyberterrorism has been conducted in the popular media, where the focus is on ratings and readership figures rather than establishing good operational definitions of new terms.
- The term is subject to chronic misuse and overuse and since 9/11, in particular, has become a buzzword that can mean radically different things to different people.
- It has become common when dealing with computers and the Internet to create new words by placing the handle *cyber*, *computer*, or *information* before another word. This may appear to denote a completely new phenomenon, but often it does not and confusion ensues.
- Finally, there is a lack of an agreed-upon definition of terrorism more generally.[19]

This does not mean that no acceptable definitions of cyberterrorism have been put forward. On the contrary, there are a number of well thought out definitions of the term available, and these are discussed below. One of the most accessible sound bites defining cyberterrorism is "hacking with a body count."[20] However, as no single, globally accepted definition of classical political terrorism exists, there is also no single definition of cyberterrorism.

Mark M. Pollitt's article "Cyberterrorism: Fact or Fancy?" published in *Computer Fraud and Security* in 1998, made a significant contribution to the definition of cyberterrorism. Pollitt points out, as many others fail to do, that the concept of cyberterrorism is composed of two elements: cyberspace and terrorism. Cyberspace may be conceived of as "that place in which computer programs function and data moves."[21] Cyberspace as a term has its origins in science fiction writing. It first appeared in William Gibson's 1984 novel *Neuromancer*, which featured a world called cyberspace, after Cyber, the most powerful computer.[22] Terrorism is a less easily defined term. In fact, most scholarly texts devoted to the study of terrorism contain a section, chapter, or chapters devoted to a discussion of how difficult it is to define the term.[23] In his paper Pollitt employs the definition of terrorism contained in Title 22 of the United States Code, Section 2656f(d). That statute contains the following definition:

The term "terrorism" means premeditated, politically motivated violence perpetrated against non-combatant targets by sub-national groups or clandestine agents, usually intended to influence an audience.

Pollitt combines Collin's definition of cyberspace and the U.S. Department of State's definition of terrorism to produce a narrowly drawn working definition of cyberterrorism as follows:

Cyberterrorism is the premeditated, politically motivated attack against information, computer systems, computer programs, and data, which result in violence against non-combatant targets by sub-national groups or clandestine agents.[24]

A similar definition of cyberterrorism has been put forward by Dorothy Denning in numerous articles and interviews, and in her testimony on the subject before the United States Congress's House Armed Services Committee. According to Denning:

Cyberterrorism is the convergence of cyberspace and terrorism. It refers to unlawful attacks and threats of attacks against computers, networks, and the information stored therein when done to intimidate or coerce a government or its people in furtherance of political or social objectives. Further, to qualify as cyberterrorism, an attack should result in violence against persons or property, or at least cause enough harm to generate fear. Attacks that lead to death or bodily injury, explosions, or severe economic loss would be examples. Serious attacks against critical infrastructures could be acts of cyberterrorism, depending on their impact. Attacks that disrupt nonessential services or that are mainly a costly nuisance would not.[25]

Pollitt and Denning are two of only a very small number of authors to recognize and make explicit the way in which the word cyberterrorism is meaningless in and of itself and that it is only the relational elements of which the word is composed that imbue it with meaning.[26] A majority of authors appear to overlook this connection. In fact, numerous authors of articles dealing explicitly with cyberterrorism provide no definition of their object of study at all.[27]

Utilizing the definitions provided by Denning and Pollitt, the attack on the Webcam of the Wise County Circuit Court does not qualify as cyberterrorism, nor do any of the other "cyberterrorist attacks" outlined earlier. It's hardly surprising; the inflation of the concept of cyberterrorism may increase newspaper circulation, but is ultimately not in the public interest. Despite this, many scholars (and others) have suggested adopting broader definitions of the term. Many authors do this implicitly by falling into the trap of either conflating hacking and cyberterrorism or confusing cybercrime with cyberterrorism, while a number of authors fall into both of these traps. Such missteps are less arbitrary than they may first appear however, as two important academic contributions explicitly allow for such a broadening of the definition of cyberterrorism.

Virtual Violence

Traditional terrorism generally involves violence or threats of violence. However, despite the prevalent portrayal of traditional violence in virtual environments (e.g in computer games), "cyber violence" is still very much an undefined activity. It is accepted, for example, that the destruction of another's computer with a hammer is a violent act. But should destruction of the data contained in that machine, whether by the introduction of a virus or some other technological means, also be considered violence?[28] This question goes right to the heart of the definition of cyberterrorism.

In a seminal article, published in the journal *Terrorism and Political Violence* in 1997, Matthew Devost, Brian Houghton, and Neal Pollard defined cyberterrorism, or information terrorism as they refer to it, as "the intentional abuse of a digital information system, network, or component toward an end that supports or facilitates a terrorist campaign or action."[29] They conceive of cyberterrorism as "the nexus between criminal information system fraud or abuse, and the physical violence of terrorism."[30] This allows for attacks that would not necessarily result in violence against humans—although it might incite fear—to be characterized as terrorist. This is problematic because, although there is no single accepted definition of terrorism, more than 80 percent of scholars agree that the it has two integral components: the use of force or violence and a political motivation.[31] Indeed, most domestic laws define classical or political terrorism as requiring violence, the threat of violence, or the taking of human life for political or ideological ends. Devost, Houghton, and Pollard are aware of this, but wish to allow for the inclusion of pure information system abuse, that which does not employ or result in physical violence, as a possible new facet of terrorism nonetheless.[32]

Nelson, et al.'s reasoning as to why disruption, as opposed to destruction, of information infrastructures ought to fall into the category of cyberterrorism is quite different:

Despite claims to the contrary, cyberterrorism has only a limited ability to produce the violent effects associated with traditional terrorist acts. Therefore, to consider malicious activity in cyberspace terrorism, it is necessary to extend existing definitions of terrorism to include the destruction of digital property. The acceptance of property destruction as terrorism allows this malicious activity, when combined with the necessary motivations, to be defined as cyberterror.[33]

As we have seen, Mark Pollitt employs the State Department's definition of terrorism to construct his definition of cyberterrorism. Neither the State Department definition, nor Pollitt's, specifically identifies actions taken against property as terrorism. According to Nelson, et al., however, in practice the Title 22 definition "clearly includes the destruction of property as terrorism when the other conditions for terrorism are satisfied (premeditated, politically motivated, etc.)."[34] In addition, the FBI definition of terrorism explicitly includes acts against property. However, Nelson, et al. point out that both the State Department and FBI definitions are subsumed by the Department of Defense definition contained in regulation O-2000.12-H, which includes "malicious property destruction" as a type of terrorist attack. This regulation also addresses destruction at the level of binary code, which it specifically refers to under the use of special weapons.

Use of sophisticated computer viruses introduced into computer-controlled systems for banking, information, communications, life support, and manufacturing could result in massive disruption of highly organized, technological societies. Depending on the scope, magnitude, and intensity of such disruptions, the populations of affected societies could demand governmental concessions to those responsible for unleashing viruses. Such a chain of events would be consistent with contemporary definitions of terrorist acts."[35] Taking the above into account, Nelson, et al. define cyberterrorism as follows: "Cyberterrorism is the unlawful destruction or disruption of digital property to intimidate or coerce governments or societies in the pursuit of goals that are political, religious, or ideological."[36] The problem is that this definition massively extends the terrorist remit by removing the requirement for violence resulting in death and/or serious destruction from the definition and lowering the threshold to "disruption of digital property."

A related problem is that although Nelson, et al. are quite precise in their categorizations and repeatedly stress that the other conditions necessary for an act to be identified as terrorist must be satisfied (e.g. premeditation, political motivation, etc.) before disruptive cyber attacks may be classified as cyberterrorism, others are less circumspect. Israel's former science minister, Michael Eitan, has deemed "sabotage over the Internet" as cyberterrorism.[37] According to the Japanese government cyberterrorism aims at "seriously affecting information systems of private companies and government ministries and agencies by gaining illegal access to their computer networks and destroying data."[38] A report by the Moscow-based ITAR-TASS news agency states that, in Russia, cyberterrorism is perceived as "the use of computer technologies for terrorist purposes."[39] Yael Shahar,

webmaster at the International Policy Institute for Counterterrorism (ICT), located in Herzliya, Israel, differentiates between a number of different types of
what he prefers to call information terrorism: electronic warfare occurs when hardware is the target, psychological warfare is the goal of inflammatory content, and
it is only hacker warfare, according to Shahar, that degenerates into cyberterrorism.[40]

Hacking versus Cyberterrorism

"Hacking" is the term used to describe unauthorized access to or use of a computer system. The term "hacktivism" is composed of the words "hacking" and
"activism" and is the handle used to describe politically motivated hacking. "Cracking" refers to hacking with a criminal intent; the term is composed of the words
"criminal" and "hacking." In a majority of both media reports and academic analyses of cyberterrorism, one or other of these terms—hacking, hacktivism, cracking—or the activities associated with them are equated with or identified as variants
of cyberterrorism.

Hackers have many different motives. Many hackers work on gaining entry
to systems for the challenge it poses. Others are seeking to educate themselves
about systems. Some state that they search for security holes to notify system
administrators while others perform intrusions to gain recognition from their peers.
Hacktivists are politically motivated; they use their knowledge of computer systems to engage in disruptive activities on the Internet in the hopes of drawing
attention to a political cause. These disruptions take many different forms, from
denial of service (DoS) attacks that tie up websites and other servers, to posting
electronic graffiti on the home pages of government and corporate websites, to the
theft and publication of private information on the Internet. Crackers hack with
the intent of stealing, altering data, or engaging in other malicious damage.[41] A
significant amount of cracking is carried out against businesses by former employees.

The term hacker was originally applied to those early pioneers in computer
programming who continually reworked and refined programs. This progressed,
R.G. Sparague explains, to the "displaying of feats of ingenuity and cleverness, in
a productive manner, involving the use of computer systems."[42] Gaining unauthorized access to computer networks was one way of displaying such expertise. This
original generation of hackers developed a code of practice, which has come to be
known as the Hacker Ethic. It was premised on two principles, namely the free
sharing of information and a prohibition against harming, altering, or destroying
any information that was discovered through this activity. Over the course of time,
however, "a new generation appropriated the word 'hacker' and with help from
the press, used it to define itself as password pirates and electronic burglars. With
that the public perceptions of hackers changed. Hackers were no longer seen as
benign explorers but malicious intruders."[43] As a result, the classical computer
hacker—bright teenagers and young adults who spend long hours in front of their
computer screens—is now the "cyberpunk."

Hackers as Terrorists

Much has been made of the similarities between profiles of terrorists and those of hackers. Both groups tend to be composed primarily of young, disaffected, males.[44] In the case of computer hackers, a distinct psychological discourse branding them the product of a pathological addiction to computers has emerged. In fact, a large number of hackers who have been tried before the criminal courts for their exploits have successfully used mental disturbance as a mitigating factor in their defense, and have thus received probation with counseling instead of jail time.[45]

Hackers are commonly depicted as socially isolated and lacking in communication skills. Their alleged anger at authority is said to reduce the likelihood of their dealing with these frustrations directly and constructively. In addition, the flexibility of their ethical systems; lack of loyalty to individuals, institutions, and countries; and lack of empathy for others are said to reduce inhibitions against potentially damaging acts. At the same time, their description as lonely, socially naive, and egotistical appears to make them vulnerable to manipulation and exploitation.[46]

Some hackers have demonstrated a willingness to sell their skills to outsiders. The most famous example is the Hanover Hackers case. In 1986, a group of hackers in Hanover, Germany, began selling information they obtained through unlawfully accessing the computer systems of various Departments of Energy and Defense, a number of defense contractors, and the U.S. Space Agency NASA, to the Soviet KGB. Their activities were discovered in 1988, but nearly two years elapsed before the group were finally identified and apprehended.[47] During the first Gulf War, between April 1990 and May 1991, a group of Dutch hackers succeeded in accessing U.S. Army, Navy, and Air Force systems. They sought to sell their skills and sensitive information they had obtained to Iraq, but were apprehended by police in the Netherlands.[48]

According to Gregory Rattray, a majority of the analyses of hackers-for-hire— who he calls "cybersurrogates" for terrorism—generally stress the ease and advantages of such outsourcing.[49] These analysts presume that terrorist groups will be able to easily contact hackers-for-hire, while keeping their direct involvement hidden through the use of proxies. The hackers could then be employed to reconnoiter enemy information systems to identify targets and methods of access. Furthermore, it is posited that if hacker groups could be employed to actually commit acts of cyberterrorism, terrorist groups would improve their ability to avoid culpability or blame altogether. Rattray does flag the important risks and disadvantages to such schemes, however. First, seeking to employ hackers to commit acts not just of disruption, but also of significant destruction that may involve killing people would in all likelihood prove considerably more difficult than buying information for the purposes of intelligence gathering. Second, simply contacting, never mind employing, would-be hackers-for-hire would subject terrorists to significant operational security risks. Third, terrorist organizations run the risk of cybersurrogates being turned into double agents by hostile governments. All three scenarios, Rattray admits, weigh heavily against the employment of cybersurrogacy as a strategy.[49]

And these are not the only risks facing terrorists who plan to employ information technology (IT) to carry out attacks. In their paper "The IW Threat from Sub-State Groups: An Interdisciplinary Approach", Andrew Rathmell, Richard Overill, Lorenzo Valeri, and John Gearson point out that should the terrorists themselves lack sufficient computer expertise, there is the likelihood that they would recruit hackers who would prove insufficiently skilled to carry out the planned attacks. In addition, these authors concur with Rattray that there is a strong case to be made for such hackers changing sides. This is because the primary motive of the hacker-for-hire is financial gain thus, given sufficient monetary inducement, such individuals are unlikely to object to reporting to other than their original "employer."[50]

David Tucker also has some interesting insights into the hacker-for-hire scenario, based on a simulation in which he took part involving a hacker and members of a number of terrorist organizations. Of the terrorists who took part in the conference/simulation that Tucker attended, one was a member of the Palestinian Liberation Organization (PLO), two were members of Basque Fatherland and Liberty (ETA), one from the Liberation Tigers of Tamil Eelam (LTTE), and one from the Revolutionary Armed Forces of Colombia (FARC). Tucker foresees potential organizational problems for any hacker-terrorist collaboration. He points out that on those occasions when hackers aren't acting alone, they operate in flat, open-ended associations. This is the opposite of many terrorist groups, which are closed hierarchical organizations. There is certainly the potential for clashes between these different organizational styles, developed in different operating environments and derived from different psychological needs. Tucker reports that a former member of ETA who was involved in the simulation repeatedly stressed the need to belong and the strength of attachment to the group as characteristic of members of clandestine organizations.[51] This is not a character trait typically associated with hackers. In fact, in the simulation in which Tucker took part, the hacker and the terrorists involved disagreed over tactics and had difficulty communicating. Eventually, these difficulties became so great that it resulted in a breakdown in the simulation group. The hacker and the terrorists were simply not able to work together. Tucker observes that if the breakdown can be generalized, it would have obvious consequences for hacker-terrorist collaboration.[52]

The only likely scenario, given the above, is cyber attacks carried out by terrorists with hacking skills.[53] This is not impossible. "The current trend towards easier-to-use hacking tools indicates that this hurdle will not be as high in the future as it is today, even as it is significantly lower today than it was two years ago."[54] According to William Church, a former U.S. Army Intelligence Officer:

If you look at the Irish Republican Army, which was probably the closest before they made peace, they were on the verge of it. They had computer-oriented cells. They could have done it. They were already attacking the infrastructure by placing real or phony bombs in electric plants, to see if they could turn off the lights in London. But they were still liking the feel of physical weapons, and trusting them.[55]

Terrorists are generally conservative in the adoption of new tools and tactics.[56] Factors influencing the adoption of some new tool or technology would include: the terrorist group's knowledge and understanding of the tool, and their trust in it. Terrorists generally only put their trust in those tools that they have designed and built themselves, have experimented with, and know from experience will work. It's for this reason that weapons and tools generally proliferate from states to terrorists.[57]

Kevin O'Brien and Joseph Nusbaum suggest that intelligence agencies should utilize online chat forums, hacker websites, and similar avenues to gather intelligence on contemporary asymmetric threats. They suggest that most hackers possess a large degree of hubris with regards to their hacking knowledge and abilities as a result of which such "threat-savvy users" could be coaxed into revealing vulnerabilities they had discovered on the Internet, as well as boasting about their own abilities and exploits.[58] David Smith, the man responsible for transmitting the Melissa virus, helped the FBI bring down several major international hackers. Smith used a fake online identity to communicate with and track other hackers from around the world. His intelligence gathering resulted in the arrest of both Jan DeWit, the author of the Anna Kournikova virus, and Simon Vallor, the author of the Gokar virus.[59] This policing tactic is endorsed by Kevin Soo Hoo, Seymour Goodman, and Lawrence Greenberg:

> Foreign bases of operation might be useful for intelligence-gathering activities, but again, they are not required for IT-enabled terrorism . . . [I]nformation about various systems' vulnerabilities is often shared online between hackers on computer bulletin boards, websites, news groups, and other forms of electronic association, and this information can be obtained without setting foot in the target country.[60]

It seems unlikely, however, that professional hackers or cyber mercenaries would engage in the cavalier behavior described above:

> While amateur hackers receive most publicity, the real threat are the professionals or "cyber mercenaries." This term refers to highly skilled and trained products of government agencies or corporate intelligence branches that work on the open market. The Colombian drug cartels hired cyber mercenaries to install and run a sophisticated secure communications system; Amsterdam-based gangs used professional hackers to monitor and disrupt the communications and information systems of police surveillance teams.[61]

There is no evidence of such mercenaries having carried out attacks under the auspices of known terrorist organizations, however.

The U.S. Department of Justice labeled Kevin Mitnick, probably the world's most famous computer hacker, a "computer terrorist."[62] On his arraignment, Mitnick was denied access not only to computers, but also to a phone, "the judge believing that, with a phone and a whistle, Mitnick could set off a nuclear attack."[63] Before the widespread deployment of all-digital switches, one could actually hear the switching tones used to route long-distance calls. "Phreaking" is the term used to describe the art and science of cracking the phone network. Early phreakers built devices called "blue boxes" that could reproduce the tones used by phone companies, which could be used to commandeer portions of the phone network. The reference above is to an early phreaker who acquired the sobriquet "Captain Crunch" after he proved that he could generate switching tones with a plastic whistle pulled out of a box of Cap'N Crunch® cereal! But at no time did he seek to set off any nuclear device using this method. Incredulity aside, hackers are unlikely to become terrorists, because their motives are divergent. Despite the allegedly similar personality traits shared by both terrorists and present-day hackers, the fact remains that terrorism is an extreme and violent occupation, and far more aberrant than prankish hacking. Although hackers have demonstrated that they are willing to crash computer networks to cause functional paralysis and even significant financial loss, this propensity for expensive mischief is not sufficient evidence that they would be willing to jeopardize lives or even kill for a political cause.[64]

Hacktivism versus Cyberterrorism

Hacktivism grew out of hacker culture, although there was little evidence of sustained political engagement by hackers prior to the mid-1990s.[65] Many view 1998 as the year in which hacktivism really took off.[66] Probably the first incidence of hacktivism took place in 1989 when hackers with an anti-nuclear stance released a computer worm into NASA's SPAN network. The worm carried the message "Worms Against Nuclear Killers. . . . Your System has Been Officially WANKed. . . . You talk of times of peace for all, and then prepare for war." At the time, anti-nuclear protesters were seeking to stop the launch of the shuttle that carried the plutonium-fuelled Galileo probe on the first leg of its voyage to Jupiter.[67] It was in 1998 that the U.S.-based Electronic Disturbance Theater (EDT) first employed its FloodNet software in an effort to crash various Mexican government websites to protest the treatment of indigenous peoples in Chiapas and support the actions of the Zapatista rebels. FloodNet is a Java applet that, once the launching page has been accessed, repeatedly loads pages from targeted networks. If enough people participate in a FloodNet attack (i.e., access the launching page at a given date and window of time), the targeted computer will be brought to a halt, bombarded by too many commands for it to process. More than eight thousand people participated in this, one of the first digital sit-ins. Probably the very first such demonstration was carried out against the French government. On 21 December 1995, a group called Strano Network launched a one-hour 'Net'Strike' attack against websites operated by various French government agencies. It was reported that some of the sites were inaccessible during that time.[68] It was also in 1998 that a young British hacker known as 'JF,' entered about three hundred websites and replaced their home pages with anti-nuclear text and imagery. At that time, JF's hack was the biggest political hack of its kind. "Hacktions" also took place in Australia, China, India, Portugal, Sweden, and elsewhere in the same year.[69] Michael Vatis, one-time Director of the FBI's National Infrastructure Protection Center (NIPC), has labeled such acts as cyberterrorism.[70]

Tim Jordan identifies two different types of hacktivism: Mass Virtual Direct Action (MVDA) and Individual Virtual Direct Action (IVDA). According to Jordan:

> Mass Virtual Direct Action involves the simultaneous use, by many people, of the Internet to create electronic civil disobedience. It is named partly in homage to the dominant form of offline protest during the 1990s, non-violent direct action or NVDA.[71]

The FloodNet attack on the Mexican government websites described above was an example of MVDA, as was the action against the 1999 World Trade Organization (WTO) conference in Seattle. The organizers of the latter event, the UK-based Electrohippies, estimated that over four hundred and fifty thousand people participated in their sit-in on the WTO website. In contrast to MVDA, IVDA utilizes classical hacker/cracker techniques and actions for attacking computer systems, but employs them for explicitly political purposes. Jordan makes the point that the name IVDA does not mean the actions are necessarily undertaken by those acting alone, but instead that the nature of such actions means that they must be taken by individuals (i.e., they in no way rely on mass action), although they may be taken by many individuals acting in concert.[72] JF's anti-nuclear protest described above was an example of IVDA, which generally consists of infiltration of targeted networks and semiotic attacks (i.e., website defacements). The major difference between MVDA and IVDA, apart from those already described, is that MVDA activists rarely seek to hide their identities— through the use of pseudonyms (handles), for example—or cover their tracks. Advocates of MVDA seek to gather together large groups of people to take part in hacktions and thus to inspire public debate and discussion, and maintain that they have a right to protest even if some of those protests are illegal or bordering on illegal. Many of those using IVDA, on the other hand, act alone and prefer to remain anonymous, which raises issues of representation, authenticity, and so on.[73] Finally, there are also differences between those hacktivists who are devoted to the classical hacking ideal of free flow of information and therefore view DoS attacks as wrong in principle and those who view MVDA as both direct non-violent action and important symbolic protest.[74]

It is the disruptive nature of hacktions that distinguishes this form of "direct action Internet politics" or "electronic civil disobedience" from other forms of online political activism. E-mail petitions, political websites, discussion lists, and a vast array of other electronic tools have been widely adopted as recruitment, organizing, lobbying, and communicating techniques by social movements and political organizations of all sorts. Stefan Wray has described this type of use of the Internet by political activists as "computerized activism."[75] The hacktivist movement is different, because it does not view the Internet simply as a channel for communication, but also crucially as a site for action. It is a movement united by its common method as opposed to its common purpose.[76] Those political causes that have attracted hacktivist activity range from campaigns against globalization, restrictions on encryption technology, and political repression in Latin America to abortion, the spread of electronic surveillance techniques, and environmental protection. Hacktivists are, therefore, arrayed across a far wider political spectrum than the techno-libertarian agenda with which committed "netizens," including the hacker fraternity, are often identified.

Hacktivists, although they use the Internet as a site for political action, are not cyberterrorists. They view themselves as heirs to those who employ the tactics of trespass and blockade in the realm of real-world protest. They are, for the most part, engaged in disruption not destruction. According to Carmin Karasic, the software engineer who designed the FloodNet program: "This isn't cyberterrorism. It's more like conceptual art."[77] Ronald Deibert is correct when he states that, while Dorothy Denning's definition of cyberterrorism is accurate and illuminating, her portrayal of hacktivism in her article "Activism, Hacktivism, Cyberterrorism" is misleading. It employs the typical practice of conflating hacking with criminal activity. This is an association that not only ignores the history of hacking, but what many view as the positive potential of hacking as a tool for legitimate citizen activism.[78] Denning appears to have adopted a more moderate position in her later work;[79] Michael Vatis, on the other hand, continues to view hacktivists as perpetrators of low-level cyberterrorism.

Cyber Crime versus Cyberterrorism

The issue of computer crime was first raised in the 1960s, when it was realized that computers could easily be employed to commit a variety of frauds. Cyber crime is a more recent phenomenon, which was enabled with the introduction of the modem and the ability to remotely access computer systems, the explosion of e-commerce, and the resultant increase in financial transactions taking place via the Internet. Attempts to conflate cyberterrorism and cyber crime were inevitable. A UN manual on IT-related crime recognizes that, even after several years of debate among experts on just what constitutes cyber crime and cyberterrorism, "there is no internationally recognized definition of those terms."[80] Nevertheless, it is clear that while cyberterrorism and cyber crime both employ information technology, their motives and goals do not coincide. Cyber criminals have financial gain as their primary motive.

[W]e have entered a new age of computer crime. With the rise of E-commerce, the development of the Net as a commercial entity, and unparalleled media attention, the profit motive for computer crime has entered the stratosphere. Recently, Janet Reno (former Attorney General of the United States) dubbed it a "huge growth industry." She's probably not wrong. What Reno and other agents of law enforcement are talking about is not hacking, it is crime. It is the kind of crime where people are hurt, money is stolen, fraud is committed, and criminals make money. It is not the grey area of electronic trespass or rearranged Web pages. It is not the world of electronic civil disobedience and "hacktivism" . . . In short, it [is] about money, and that makes it a different kind of crime.[81]

Areas in which individual criminals and criminal organizations have proven proficient in cyberspace include the theft of electronic funds, the theft of credit card information, extortion, and fraud.[82] Secondary to financial gain is the acquisition of information that can underpin the operations associated with making money. It is for this reason that transnational crime syndicates are probably more interested in maintaining a functioning Internet than attacking Internet infrastructures. In other words, organized crime groups view the Internet as a tool, not a target. This is because many such organizations employ the Internet—and the public telecommunications network generally—as a vehicle for intelligence gathering, fraud, extortion, and theft.[83] For example, as banks and other financial institutions increasingly rely on the Internet for their daily operations, they become more attractive targets for criminal activity. Having said that, criminal groups, such as drug traffickers, may seek to penetrate information systems to disrupt law enforcement operations or collect information on operations planned against them.[84]

This does not mean that the proceeds of cyber crime may not be used to support terrorism, but only that were this to occur it ought not to be classed as cyberterrorism per se.

Computer as Target versus Computer as Tool

In a probing article simply titled "Cyberterrorism?" , Sarah Gordon and Richard Ford draw the reader's attention to the differences between what they call "traditional cyberterrorism" and "pure cyberterrorism." According to Gordon and Ford, traditional cyberterrorism features computers as the target or the tool of attack while pure cyberterrorism is more restricted as it is limited to attacks against computers, networks, etc.[85] The authors point out that both the media and the general public favor the definition encapsulated in the term "traditional cyberterrorism" while the focus in academia is on "pure cyberterrorism." So, while conceding that Denning's—and thence Pollitt's—definition is "solid," Gordon and Ford find the definition less than comprehensive:

First, [Denning] points out that this definition is usually limited to issues where the attack is against "computers, networks, and the information stored therein," which we would argue is "pure cyberterrorism." Indeed, we believe that the true impact of her opening statement ("the convergence of terrorism and cyberspace") is realized not only when the attack is launched against computers, but when many of the other factors and abilities of the virtual world are leveraged by the terrorist in order to complete his mission, whatever that may be. Thus, only one aspect of this convergence is generally considered in any discussion of cyberterrorism—an oversight that could be costly. Second, it is very different from the definition that appears to be operationally held by the media and the public at large.[86]

A number of authors agree with Gordon and Ford that cyberterrorism should encompass any act of terrorism that utilizes "information systems or computer technology as either a *weapon* or a *target*."[87] Nelson, et al. include physical attacks upon information infrastructures in this category.[88] However, the same authors disagree with Gordon and Ford on the issue of leveraging the abilities of the virtual world to complete a terrorist mission. Gordon and Ford seek to place the latter activity squarely in the category of cyberterrorism. Nelson, et al. emphatically reject this approach. They identify two new categories into which this type of activity may be placed, "cyberterror support" and terrorist "use" of the Internet. "Cyberterror support is the unlawful use of information systems by terrorists which is not intended, by itself, to have a coercive effect on a target audience. Cyberterror support augments or enhances other terrorist acts." On the other hand, "terrorist use of information technology in their support activities does not qualify as cyberterrorism."[89]

Distinguishing Characteristics

Kent Anderson suggests a three-tiered schema for categorizing fringe activity on the Internet, wherein activities are classed as use, misuse, or offensive use. Anderson explains:

> Use is simply using the Internet/WWW to facilitate communications via e-mails and mailing lists, newsgroups and websites. In almost every case, this activity is simply free speech. . . . Misuse is when the line is crossed from expression of ideas to acts that disrupt or otherwise compromise other sites. An example of misuse is Denial –of Service (DoS) attacks against websites. In the physical world, most protests are allowed, however, [even] if the protests disrupt other functions of society such as train service or access to private property. . . . The same should be true for online activity. Offensive use is the next level of activity where actual damage or theft occurs. The physical world analogy would be a riot where property is damaged or people are injured. An example of this type of activity online is the recent attack on systems belonging to the world economic forum, where personal information of high profile individuals was stolen.[90]

Combining Anderson's schema with the definitions of cyberterrorism outlined by Pollitt and Denning, it is possible to construct a four-level scale of the uses (and abuses) of the Internet for political activism by unconventional actors, ranging from use at one end of the spectrum to cyberterrorism at the other (see Table 1). Unfortunately, such a schema has not generally been employed in the literature or in the legislative arena. This is particularly disquieting given that the vast majority of terrorist activity on the Internet is limited to use.[91]

Legislative Measures

In February 2001, the UK updated its Terrorism Act to classify "the use of or threat of action that is designed to seriously interfere with or seriously disrupt an electronic system" as an act of terrorism.[92] In fact, it will be up to police investigators to decide whether an action is to be regarded as terrorism. Online groups, human rights organizations, civil liberties campaigners, and others condemned this classification as absurd, pointing out that it placed hacktivism on a par with life-threatening acts of public intimidation.[93] Furthermore, ISPs in the UK may be legally required to monitor some customers' surfing habits if requested to do so by the police under the Regulation of Investigatory Powers Act 2000. In the wake of the events of 9/11, U.S. legislators followed suit. Previous to 9/11, if one successfully infiltrated a federal computer network, one was considered a hacker. However, following the passage of the USA Patriot Act, which authorized the granting of significant powers to law enforcement agencies to investigate and prosecute

Table 1
Typology of Cyber Activism and Cyber Attacks

Action	Definition	Source	Example
Use	Using the Internet to facilitate the expression of ideas and communication(s)	Internet users	E-mails, mailing lists, newsgroups, websites
Misuse	Using the Internet to disrupt or compromise websites or infrastructure	Hackers, Hacktivists	Denial-of-Service (DoS) attacks
Offensive Use	Using the Internet to cause damage or engage in theft	Crackers	Stealing data (e.g. credit card details)
Cyberterrorism	An attack carried out by terrorists via the Internet that results in violence against persons or severe economic damage	Terrorists	A terrorist group using the Internet to carry out a major assault on the New York Stock Exchange

potential threats to national security, there is the potential for hackers to be labeled cyberterrorists and, if convicted, to face up to twenty years in prison.[94] Clearly, policymakers believe that actions taken in cyberspace are qualitatively different from those taken in the "real" world.

It is not the Patriot Act, however, but the massive five-hundred-page law establishing the U.S. Department of Homeland Security that has the most to say about terrorism and the Internet. The law establishing the new department envisions a far greater role for the United States government in the securing of operating systems, hardware, and the Internet in the future. In November 2002, U.S. President Bush signed the bill creating the new department, setting in motion a process that will result in the largest reshuffle of U.S. bureaucracy since 1948. At the signing ceremony, Bush said that the "department will gather and focus all our efforts to face the challenge of cyberterrorism."[95]

The Department of Homeland Security merges five agencies that currently share responsibility for critical infrastructure protection in the United States: the FBI's National Infrastructure Protection Center (NIPC), the Defense Department's National Communications System, the Commerce Department's Critical Infrastructure Office, the Department of Energy's analysis center, and the Federal Computer Incident Response Center. The new law also creates a Directorate for Information Analysis and Infrastructure Protection whose task it will be to analyze vulnerabilities in systems including the Internet, telephone networks, and other critical infrastructures, and orders the establishment of a "comprehensive national plan for securing the key resources and critical infrastructure of the United States" including information technology, financial networks, and satellites. Further, the law dictates a maximum sentence of life-imprisonment without parole for those who deliberately transmit a program, information, code, or command that impairs the performance of a computer or modifies its data without authorization, "if the offender knowingly or recklessly causes or attempts to cause death." In addition, the law allocates $500 million for research into new technologies, is charged with funding the creation of tools to help state and local law enforcement agencies thwart computer crime, and classifies certain activities as new computer crimes.[96]

Concluding Thoughts

In the space of thirty years, the Internet has metamorphosed from a U.S. Department of Defense command-and-control network consisting of less than one hundred computers to a network that crisscrosses the globe. Today the Internet is made up of tens of thousands of nodes (i.e., linkage points) with more than 105 million hosts spanning more than two hundred countries. With an estimated population of regular users of more than six hundred million people, the Internet has become a near-ubiquitous presence in many world regions. That ubiquity is due in large part to the release in 1991 of the World Wide Web. In 1993 the Web consisted of a mere 130 sites, by century's end it boasted more than one billion. In the

Western world, in particular, the Internet has been extensively integrated into the economy, the military, and society as a whole. As a result, many people now believe that it is possible for people to die as a direct result of a cyberterrorist attack and that such an attack is imminent.

On Wednesday morning, 12 September 2001, you could still visit a website that integrated three of the wonders of modern technology: the Internet, digital video, and the World Trade Center. The site allowed Internet users worldwide to appreciate what millions of tourists have delighted in since Minoru Yamasaki's architectural wonder was completed in 1973: the glorious forty-five-mile view from the top of the WTC towers. According to journalists, the caption on the site still read "Real-Time Hudson River View from World Trade Center." In the square above was deep black nothingness. The terrorists hadn't taken down the Internet; they had taken down the towers. "Whereas hacktivism is real and widespread, cyberterrorism exists only in theory. Terrorist groups are using the Internet, but they still prefer bombs to bytes as a means of inciting terror," wrote Dorothy Denning just weeks before the September attacks.[97] Terrorist use of the Internet has been largely ignored, however, in favor of the more headline-grabbing "cyberterrorism."

Richard Clarke, former White House special adviser for Cyberspace Security, has said that he prefers not to use the term cyberterrorism, but instead favors use of the term "information security" or "cyberspace security." This is because, Clarke has stated, most terrorist groups have not engaged in information warfare (read "cyberterrorism"). Instead, he admits, terrorist groups have at this stage only used the Internet for propaganda, communications, and fundraising. In a similar vein, Michael Vatis, former head of the U.S. National Infrastructure Protection Center (NIPC), has stated that "Terrorists are already using technology for sophisticated communications and fund-raising activities. As yet we haven't seen computers being used by these groups as weapons to any significant degree, but this will probably happen in the future."[98] According to a 2001 study, 75 percent of Internet users worldwide agree; they believe that "cyberterrorists" will "soon inflict massive casualties on innocent lives by attacking corporate and governmental computer networks." The survey, conducted in nineteen major cities around the world, found that 45 percent of respondents agreed completely that "computer terrorism will be a growing problem," and another 35 percent agreed somewhat with the same statement.[99] The problem certainly can't shrink much, hovering as it does at zero cyberterrorism incidents per year. That's not to say that cyberterrorism cannot happen or will not happen, but that, contrary to popular perception, it has not happened yet.

CHAPTER 6

Information Operations Education: Lessons Learned from Information Assurance

Dr. Corey Schou, Director, National Information Assurance Training and Education Center (NIATEC), Idaho State University
Dr. Dan Kuehl, National Defense University, Leigh Armistead, Edith Cowan University

I nformation operations (IO) and information assurance (IA) are two sides of the same coin. In the United States, there has been a concerted effort by industry and government to stimulate the civilian academic community to take more interest in IA. However, the same emphasis has not been placed on the universities with respect to IO, and its academic development has lagged. Information Operations as a formal term dates back seven years to the publication of Joint Publication 3-13 (JP 3-13) by the United States Department of Defense (DoD). Preceded by other theories and concepts such as command and control warfare (C2W) and information warfare (IW), the basic idea is that information, like military, diplomatic, and economic power, is a force that can be utilized alongside of and in conjunction with those other elements of national power to support military and national security strategy. This construct has evolved consistently since the end of the first Gulf War, and IO is now a recognized element and area of study with the American military forces, with well over fifty courses in existence. Yet the same cannot be said of the academic community within the United States. To date, there are no IO degree programs or certificate granting institutions concentrating in this field. While a large number of universities have developed more specialized IA or computer security programs, the broader aspects of information operations are still not seen as worthy of study by the academic community. Yet in Great Britain and Australia this is not the case, where a number of colleges have built post-graduate curricula emphasizing IO.

Key Questions
There are several fundamental questions that need to be addressed.

- Why does this dichotomy exist?
- Why is there such a broad consensus on the need for this kind of education

and training by the United States military, when the academic community does not offer equivalent education or classes?

- Why are other nations moving ahead of America with respect to IO curricula and should there be more development of these types of programs in universities in the United States?
- What can the United States do to solve the problem?

From the lessons that were learned over the last few years by the IA community, this paper focuses on answering the latter question. To do this, we will first examine the evolution and background of information operations (IO) within the United States. From there, we will propose a nine-point program to establish an integrated academic infrastructure dedicated to providing the education and training required to support the use of IO to protect critical elements of the national information infrastructure.

Background

It is widely acknowledged that we are living in a new era, an information era, where new technology and the merging of different functional areas, like telecommunications and computing— a field known as "telematics"—are setting up a networking capability that is transforming the way we live. Those who can control the flow of information can control power. Yet in this modern era, the government can no longer monopolize or control information but can only coordinate its flow. We now live in the age of networks and international organizations. Nation-states are losing power to hybrid organizations that rely on other types of affinity such as religion or ethnic backgrounds within a networked communication architecture.. This change in behavior is also shifting the way that we look at information, which alters our understanding of power within the world political structure.[1] Interconnectivity and access are the two key pillars of these new organizations. Truth and guarded openness have become the approach used both in the private and government sector to conduct business, and globalization has diminished the importance of borders.[2] Key features of these changes include:

- Open, omnidirectional communication links where speed is everything,
- A common language for almost any type of information—the bit and byte
- Information in almost limitless volume,
- Little to no censorship: the individual will control his own information flow,
- Truth and quality will surface but perhaps not initially.

While some key decision-makers may understand and appreciate this revolutionary change, many do not. There is a gap between the military and civilian leadership. Correcting this requires an education process capable of delivering a tailored package to individuals, whether members of the armed forces, senior government officials, or corporate leaders, that conveys the importance of the power of information. Yet, while there is formal training on the concept and importance of IO within the services, joint commands, and government agencies, this same emphasis does not exist in American universities. While there is tremendous understanding and acknowledgement that information is the key to the future of warfare and business, this has not spawned the development of dedicated civilian academic programs at the graduate and post-graduate level. For it is not just officers or a small group of military personnel who need to understand these concepts but, as will be demonstrated throughout this paper, the general public needs to understand and use the power of information.

The Evolution of Information Operations Theory

The development of information operations as a major military doctrine in the USG is a relatively new phenomenon, and much of that critical thinking began in the early 1980s.[3] With the demise of the Soviet threat to the continental United States and the shift from bipolar to multipolar political scenarios, the American force structure and military doctrine has evolved dramatically. Moving away from garrisoned forces designed to combat a monolithic enemy, the emphasis today is on global operations against terrorist and ideological foes. However, the biggest shift in military doctrine has been shaped not by politics, but by the huge technological changes that evolved over the last ten to fifteen years. The advances in computers, software, telecommunications, and networks, have revolutionized the way the United States conducts military operations and have given it premier armed forces.

From these experiences and other missions since the end of the cold war, the increase in the value of information is one of the most important lessons learned. Time after time commanders have realized that the side possessing and controlling the most information, with the ability to accurately manipulate and conduct an influence campaign is to be victorious. Much of this thinking in the American armed forces began immediately after the fall of the Soviet Union, when strategic planners at the Joint Chiefs of Staff laid down a series of new war fighting strategies. Highly classified, excerpts and briefs laid out the case for using information as a warfare tool.[4] These documents started a dialogue on IW within the DoD. Unfortunately, the classification of such information prohibited a more general doctrinal exchange within government circles. Consequently, decision-makers desired a broader explanation of the power and use of information in this new environment.

The need for a strategy that incorporated these technological revolutions resulted in the new concept of command and control warfare (C2W). Officially released as the Chairman of the Joint Chiefs of Staff Memorandum of Policy 30 (CJCS MOP 30) Command and Control Warfare on 8 March 1993, this document laid out for the first time in an unclassified format the interaction of the different C2W disciplines, which gave the fighters the IW advantage. Military intelligence was used to support five pillars of offensive and defensive C2W. These five pillars are destruction, deception, psychological operations (PYSOP), operations security (OPSEC), and electronic warfare (EW).

Individually, none of these operations were new to warfare. All played prominent roles during the June 1944 Operation Overlord landing in Normandy, for
example, and examples of some of them—OPSEC, PSYOP and deception—can
be found as far back as the Old Testament. Some quarters of the military greeted
this new concept of warfare with enthusiasm, while others were wary of any new
doctrinal developments. However, the ability to integrate these different military
disciplines to conduct nodal analysis of enemy command and control targets was
also highly lauded as a great improvement. Many units and all four services developed C2W cells and began training in this new doctrine throughout the mid-1990s.
But there was still a conflict between the CJCS MOP 30 and the DoDD TS3600.1
doctrine, since IW represented a much broader attempt to tackle the issue of information as a force multiplier, while C2W was more narrowly defined in terms of
the five pillars. Additionally, the fact that the United States was writing strategy to
conduct operations against nations in peacetime was considered very risky; therefore IW remained highly classified throughout much of the 1990s.

Yet the United States military recognized the organizational benefits of conducting this type of warfare in the information age and therefore, even though
doctrine was still in the formative stage, organizational changes began to occur in
the early 1990s. The Joint Electronic Warfare Center at Kelly AFB in San Antonio, Texas, was renamed the Joint Command and Control Warfare Center in 1993.
It would later be renamed the Joint Information Operations Center (JIOC) in October 1999. Likewise the uniformed services created a number of other new agencies beginning in 1995, including:

- U.S. Air Force—Air Force Information Warfare Center (AFIWC)
- U.S. Army—Land Information Warfare Activity (LIWA), now known as the
  1st Information Operations Command
- U.S. Navy—Fleet Information Warfare Center (FIWC), now Navy Information Operations Command (NIOC), a subordinate command of Navy Network Warfare Command

These organizational developments fully conformed to classic revolutions
in military affairs (RMA) theory, which holds that technological change only provides the starting point, because real progress and maturation require both organizational and doctrinal adaptation.

Policy surrounding the use of information by the United States military also
continued to develop after the publication of MOP 30. The formation of IW agencies and commands in 1995 was very useful, for it not only filled voids in the
services capabilities, but also helped to resolve the conflict in the development of
information doctrine and policy within the USG. Thus in the mid-1990s, there
was a concerted push for declassification and better understanding of these
concepts within the DoD, which resulted in the 9 December 1996 publication of
DoDD S3600.1, Information Operations.By downgrading this document to secret
classification, the DoD opened the discussion of IO to a wider audience. In a related effort, the Defense Science Board also published its Report Of The Defense

Science Board Task Force On Information Warfare - Defense (IW-D)[5] in November 1996. Together these two documents attempted to clarify the differences between the older doctrine on C2W, and for the first time introduced the use of computer network attack (CNA) as an IO capability.

Thus, the formation of IW agencies and commands between 1995 and 1996 did more than anything else to help resolve the conflict in the development of IO doctrine and policy within the United States. However, since the DoDD S3600.1 publication was classified secret, it still limited greater discussion on the differences between IO and IW.[6] Thus the mid-to-late 1990s were also a period of early experimentation. The military conducted a number of exercises elevating the awareness of information operations within the military and civilian communities. The CNA exercises conducted in 1996 and 1997 were also particularly effective and drew attention to the fact that the DoD was vulnerable to this type of attack. However, there were still questions regarding IO definitions that would not be fully addressed until the release of the seminal publication, Joint Publication 3-13, Joint Doctrine for Information Operations on 9 October 1998. For the first time, the DoD released an unclassified document that widely disseminated the principles involved in conducting IO and using the power of information. Since influence-campaigns are often conducted long before the beginning of active hostilities, the White House, DoD, and State Department realized they needed better coordination and integration of information efforts in order to be successful. This interaction between federal agencies within the executive branch also brought about a renewed emphasis on the IO organizational structure within the United States and the need for interagency training and education capabilities.

Development of DoD Information Operations Training Courses

From the newly developed principles of IO and IW came new courses and schools that were developed to teach tactics and to explore the deeper strategic issues that swirled around these new technologies and approaches to warfare. The National Defense University (NDU) created a School of Information Warfare and Strategy in 1994. It featured a full ten-month curriculum designed to immerse a small group of senior students—equivalent to those attending NDU's National War College (NWC) or Industrial College of the Armed Forces (ICAF)—in the academic theory of IW. Held for two years, NDU graduated sixteen students the first year and thirty-two the second.

The ten-month program became the basis for the current NDU effort, the Information Strategies Concentration Program (ISCP), which is an electives-based course of study available to interested National War College and ICAF students. Offered as a compliment to and expansion of their core curricula, the ISCP is built around three key aspects of information as an element of power. Since its initial group of students in 1997, ISCP has produced nearly six hundred NWC and ICAF graduates who are better prepared to shape and execute national security strategy in the Information Age. Unfortunately, this represents less than 20 percent of the total NWC/ICAF graduates during that period, so that the great majority of the graduates of the DoD's premier joint senior educational schools received virtually no exposure to IO and information as an element of national power. The DoD does a masterful job of preparing "information warriors," defined here as expert practitioners of one of IO's core competencies, as delineated by the Information Operations Roadmap published in 2003 by the Department of Defense. Thanks to the efforts of Joint Forces Staff College (JFSC) and the Naval Postgraduate School, the DoD is also producing "information operations planners," those officers who can synchronize and synergize the various elements of IO into a coherent plan and then integrate that into a theater campaign plan. Where the DoD falls short, however, is the development of "information strategists," senior leaders who can integrate the information component of power with the other elements of national power to produce national security strategies that develop and employ all of the combined elements of national power. To these authors' knowledge NDU is the only place where this is done on a systematic basis.

Yet NDU is by no means the only IO curriculum. However, among the multitude of IO courses developed by the U.S. military, there is a problem rooted in the lack of standards and common learning objectives upon which to base goals. These different curricula are also based on different theories (service and agency), different skill levels of user (beginners to advanced), rank/grade of the audience (enlisted to flag/general officer), as well as different foci (strategic, operational, and tactical). So it should not be surprising that there are more than fifty IO courses in existence today, taught by a variety of organizations and commands, with little to no interaction or integration. For example, one cannot obtain IO training in one service and then serve in a joint organization without additional specialized training. Additionally, there are no common goals that translate well across the American armed forces with regard to IO training and educational requirements. These and other standardization issues have thwarted the U.S. military and academia in moving toward the development of curricula emphasizing the power of information in general and IO in particular.

The authors sponsored a collaborative discussion session among British, American, and Australian academics and military officers at the 2nd Annual IO Conference hosted by IQPC in London in July 2003. During this daylong session, a tremendous amount of energy and analysis was devoted to finding ways to improve access to IO training and education capabilities across the three nations. Table 1 is a synopsis of those efforts, and reiterate what the authors have been advocating for a long time, mainly that any curriculum developed must be based on open and accessible standards and that a web or Internet based set of courseware was the best answer to deliver content globally.

While this matrix is not the sole answer to the problem, the authors believe that it may help to act as a checklist or guide to help in focusing the attention on possible solutions to these IO education and training goals. Yet there is still a large gap between the number of military-oriented and civilian-oriented IO courses.

Table 1
IO Training and Education Capabilities

IO Education and Training Goals	Means
Delivery of training must be cheap and fast	Internet
Access must be worldwide and standard	Portal
Materials must be clear, concise, authoritative, and readable	Textbook
Information battlespace	COP
Create a planning tool/checklist	Excel application
Study real world operations	Case studies
Establish common IO definitions/language	Taxonomy
Change perceptions and generate interest	Exercises
Organize parallel play/multiple courses	Interfaces
Create a worldwide IO game	Everquest?
Standard IO training material	CD-ROMs
Training must be standardized	Qualifications
Red teaming must be incorporated	VA teams

This Is Not a New Experience for The United States: The History of the Development of Information Assurance Education

The dichotomy between increased emphasis by the American military on the conduct of IO and the lack of corresponding academic programs within academia is not unprecedented. In the following discussion the authors demonstrate how government, industry, and universities worked together to solve the problem of increasing the emphasis on information assurance. Over a prolonged period, information assurance (IA) experienced the same cycle of development that IO faces today. The lessons learned from that experience are useful in filling the need to develop IO educational programs for academia.

LESSONS ONE, TWO, AND THREE: KNOW THE BACKGROUND AND THE PLAYING FIELD
In 1986 the security community began to recognize the need for improved education and training, which included academic participation in information assurance. To accomplish this end, they established training standards and a defined set
of knowledge, skills, and attributes (KSA).[7] This early work at National Information Assurance Training and Education Center (NIATEC) to develop a taxonomy
led to several industry professional standards, National Institute of Standards
(NIST) publication 800-16, and the Committee for National Security Systems
(CNSS) series of publications. These standards, developed by the National Security Agency (NSA), are now widely recognized throughout the DoD and interagency as the de facto baseline of tasks for IA across the federal bureaucracy. In
addition, the CNSS series has become widely used in academia, through NSA
sponsored IA programs and curriculum. Therefore, in taking a page from the computer security professionals, here are the first three learned-lessons, which should
be employed to stimulate interest of American universities in IO to *create a us-
able taxonomy of the field; establish training standards for the field; ensure
standards are interoperable.*
      In 1997, a call to action occurred when Dr. Gene Spafford of Purdue Univer-
sity pointed out the fact that

> We are facing a national crisis in the near term that threatens our national
> security, our individual safety, and our economic dominance. The rapid
> growth of information technology is a driving factor in this threat: we are
> relying on new and often fragile technology in critical applications. Fur-
> thermore, those applications present attractive targets to criminals, van-
> dals, and foreign adversaries.[8]

LESSON FOUR: MAKE SURE OTHERS RECOGNIZE THERE IS A PROBLEM
Spafford further dramatized the criticality by pointing out that, "Over the last five
years, approximately 5,500 PhDs in Computer Sciences and Engineering were
awarded by universities in the U.S. and Canada."[9] Only sixteen of those (an average of three per year) were awarded for security-related research at these major
centers. Worse still, only eight of those sixteen graduates were U.S. nationals, and
only three of the sixteen went into academic careers. The take away from this
experience is as follows: *make sure others recognize there is a problem.*
      With leadership from visionary government leaders, the IA community be-
gan to address the problem of training and education for a corps of professionals
through a series of critical actions that included establishing the following:

- National Information Assurance Education and Training Program (NIETP)
- Department of Homeland Security (DHS) /NSA National Centers of Aca-
  demic Excellence in Computer Security Education (CAE) administered by
  NIETP
- National Science Foundation (NSF) Scholarship for Service Program (SFS)

- NSF Capacity Building Program
- DoD IATF Scholarship Program administered by NIETP
- Colloquium for Information Systems Security Education (CISSE)
- Courseware clearing house at the National Information Assurance Training and Education Center (NIATEC)
- National Standards for training of security professionals through the Committee for National Security Standards (CNSS)

LESSON FIVE: MAKE SOMETHING HAPPEN

Together these groups make up the hub of IA activity, and a tremendous amount of activity has occurred in this hub over the last decade. An entire cadre of IA professionals has been trained. These trained professionals now occupy key and influential positions within the federal government because of the education that they received from these programs. The key component has proven to be the development of the CNSS Standards, which are grouped into six categories (4011 to 4016). Updated on a regular basis, these serve as a baseline for all the certifications and academic programs sponsored by the NSA/NIETP and IA academic community.

Methods to Increase the Amount and Quality of Information Operations Education

If the problem of developing academic interest in IO is to be solved, several steps are necessary. They can be modeled on the steps originally recommended for the information assurance discipline.[10] The first is to build capacity, for if IO is to become a civilian academic field, one must have sufficient faculty. However, producing a substantial body of new faculty requires considerable lead-time. If academic programs were to experience an immediate and significant increase in students demanding programs in information operations, they could not respond. There are few college professors trained in IO. Currently the computer science, information assurance, and information systems programs in the United States would all be unable to adequately respond to the increased demand for IO courses. Likewise, there are virtually no academic programs in the other disciplines such as psychological warfare, operations security, electronic warfare, and physical destruction. An academic development program for preparing new faculty, retraining existing faculty from other specialties, and assisting current IO faculty to remain current is essential to meeting any short-term demand. However, that will not be enough to build the overall capacity truly needed. For the long-term, it will be necessary to increase faculty in all areas of information technology, not just IA and IO. Other areas such as the social sciences, public affairs, organizational management, strategic communication, and even business will need to be included in this multidisciplinary approach. One method would be to establish a program to use the current faculty leaders in IA and IO education to educate and train emerging faculty in information operations. Provision of faculty scholarships for participants would encourage wider participation. This program will increase immediately the number and quality of information operations education offerings available to industrial and governmental interests. However to do this would probably require some sort of centralized government sponsorship, which to date for the IO community, has not been forthcoming.

Drafting Professors and Distinguished Professors

Current information operations practitioners could be encouraged to enter the professoriate by the creation of academic positions for professionally qualified individuals. In the United States, there are currently no IO curricula or graduate programs in academia. If ongoing distinguished professorships are established, it will allow academic institutions to augment salary offers to the finest current and potential information operations faculty. Larger financial incentives elsewhere currently lure these essential leaders from academia. A supplemental stipend for distinguished faculty members would provide financial incentive to encourage faculty to remain in the academic community. Retaining individuals from other disciplines as faculty would assure the continued supply of well-trained and educated individuals. Failure to retain faculty places the remainder of the information operations initiative in peril. This program requires balance. If distinguished professorships are aggregated at a small number of institutions, information operations programs will not spread widely. On the other hand, if they are spread too thinly there will be an insufficient core of faculty to establish well-balanced programs. When fully implemented, one hundred distinguished professorships could be established at twenty-five to fifty universities nationwide. Institutions should be able to use effectively two to four distinguished professorships per institution.

LESSONS SIX AND SEVEN: INDUSTRY PARTICIPATION

The role of industry cannot be overlooked in making faculty retention and development easier for the IO initiative. Industry may have a vested interest in the military industrial complex; however, a prettier name for IO in many cases is something they do care about — industrial espionage. Without the active participation of the information and security industry, the proposed educational program will not be able to meet the more general needs of the information operations community. In addition, there is a pressing need for infrastructure protection, information assurance, and information operations efforts to be integrated into the commercial workplace. Several avenues for academic-industry cooperation are:

- Donation of equipment to keep teaching and research facilities current,
- Short term loans of equipment and expertise to support teaching and research,
- Provision of matching funds for student and faculty development,
- Provision of matching funds to tie scholarships to internship,
- Placement of qualified industry personnel in classrooms, and faculty in industry, similar to the Intergovernmental Personal Act (IPA).

In addition, internships can help transfer important information and skills to business and industry. Participation in internships will alert faculty and students to emerging needs in business and industry.

LESSON EIGHT: INDUSTRY CAN MAKE IT EASIER

While being helpful during the inception of the program, industry participants also have to be careful to remember one of the objectives is to build a cadre of highly educated individuals to build the next generation of IO professionals. They cannot hire all the graduates all the time. Industrial interestes must leave some in academia to form the new and growing professoriate.

LESSON NINE: ATTRACTING AND KEEPING QUALITY STUDENTS

It is imperative to attract quality students to programs producing information operations specialists. One model that applies is the National Defense Education Act (NDEA) loan program. It was a powerful stimulant to increasing the ranks of science and engineering faculty in America's graduate schools in the 1960s. A scholarship program of this type for information technology with a special emphasis on information operations would aid in developing a cadre of graduate and undergraduate students to enter the profession. This has been applied successfully in the information assurance area. The Scholarship for Service program provides stipends and allowances to the students higher than those available from other sources. This appears to be necessary, at least initially, to attract sufficient numbers of qualified students to these programs.

LESSON TEN: UNDERGRADUATE PROGRAMS—ENTRY LEVEL EMPLOYEES

As demonstrated in information assurance initiative, an undergraduate scholarship program has the largest potential influence on the short-term education initiative. It is at the undergraduate level that students learn entry-level job skills. If there were a concerted effort to infuse information operations material into the existing undergraduate information assurance curricula, new entrants to the workforce would approach jobs with the attitude that information operations is a part of their responsibilities. Of course the CNSS approach to standards could be used to build parts of this into the curriculum it is important to remember that:

- More undergraduates can be produced with fewer resources,
- Undergraduates immediately enter the job market at a practitioner level,
- Advanced undergraduates also participate in research on campus.

LESSON ELEVEN: TRADITIONAL GRADUATE PROGRAMS— RESEARCHERS AND TEACHERS

In the absence of some form of graduate stipend program, there will be a continued dearth of individuals to become the next generation professoriate and to fill governmental and industrial needs. Production of masters and doctoral students is essential. Masters students will be able to implement and design information operations tools and systems well beyond the capability of students with baccalaureate degrees. Additionally, these individuals will be involved in teaching. PhDs are the primary source of educating those seeking degrees in information operations and both applied and basic research. They are the leaders in academic, government, and industry research and development. These scholarships, tied to ongoing research in the field, will attract and retain the additional students in advanced programs. It will produce additional qualified IO researchers, educators, and practitioners. Recipients will be able to perform additional basic research during the time spent in their educational programs. The program would need competitive graduate stipends as well as other incentives. A compelling incentive would be the retirement or reduction of significant student loan debt in exchange for actual years of service in academia or government. For highly qualified students the forgiveness might be extended for completion of a PhD rather than just an MS. There is strong support for increasing the professoriate therefore a high priority should be graduate education, including scholarships for graduate education and research funds. However, PhD's are handcrafted. The structure, programs, and faculty cannot be rapidly increased because of the mentorship involved with doctoral students.

Lesson Twelve: Activities for the Profession
Another source for IO faculty and professionals can be found in the bank of current practitioners. This requires a change in thinking in the academic community. If we invest in the academic pipeline to create IO professionals, there must be additional infrastructure to support them. This can be done through continuing professional education.

Professional Graduate Programs—Educate Current Employees
Traditional undergraduate and graduate programs alone cannot meet the need for information operations professionals. Any comprehensive solution must include ongoing professional education for the existing workforce. Yet, this professional education can interrupt the work and personal lives of these professionals by asking them to participate in a traditional on-campus experience. Internet-based education offers the solution by allowing professional students to continue their current employment without geographic relocation. Institutions recognized by the National Security Agency as National Centers of Excellence are natural partners in providing a "virtual university" offering Internet-based graduate education and other innovative and cooperative approaches for working professionals. Some of these materials are available through the NIATEC website.

Lesson 13: Professional Training and General Professional Development
As a further means of professional development in information operations, the implementation of a national strategy should develop short-term summer seminars and courses in information operations directed to current employees who need to enhance their skills in this area. Completion of a program of study could

be documented by offering a post-baccalaureate certificate. These certificate programs should focus on the following:

- Technical staff also requiring non-technical skills, such as information operations risk management skills;
- graduates not equipped in information operations through traditional IT courses; and
- the need for all staff (not only IT and IA staff) to have an understanding of information operations.

LESSON 14 – KEEP THE PROFESSIONALS PROFESSIONAL

As learned in preceding IA example, meeting the national IO needs is not a trivial feat. There is no fast and simple solution. By encouraging and increasing the capacity of current programs, there will be an immediate, small increase in flow created by accelerating the progress of students currently in the programs. Currently, the production has been increased to a few hundred a year. Experience with the IA scholarship program indicates that a *de novo* program may take as long as four to five years to produce the first individuals with baccalaureate degrees focusing on information operations. To produce individuals at the masters level takes an additional year and a half and yet an additional two to three years to produce a PhD.

The foregoing discussion provides investment solutions that initiate and rapidly build an IO educational infrastructure for the long=term national interest. These solutions involve:

- investing in undergraduate and graduate students to encourage them to enter the profession,
- investing in current faculty to keep them in academia,
- investing in converting faculty from other departments to support information operations initiatives,
- investing in research to remain on the cutting edge and advance the profession,
- aiding in the development of information operations as a recognized discipline in conjunction with information assurance,
- aiding faculty in professional development and publication of research results.

SPECIFIC ACTION PLAN

The following nine-point program would establish an integrated academic infrastructure dedicated to producing IO professionals. These future professionals will have the training and education needed to protect the critical elements of U.S. information infrastructure. Specific actions proposed include:

- Creation of a NDEA style scholarship program to encourage both undergraduate and graduate students to enter the profession. This is a partial success already with the SFS and DoD scholarships receiving initial funding to support IA programs,

- creation of distinguished professorships and associated stipends to encourage faculty both to join and to remain in the academic ranks,
- creation of joint research opportunities with government,
- creation of mechanisms to maintain currency of teaching and research facilities,
- encouragement of government, industry, and academic personnel interchanges,
- encouragement of joint academic-industry research consortia to address current needs,
- creation of an information operations training program to increase the number of faculty teaching and researching in the area,
- creation of joint education and training programs to keep practitioners current,
- encouragement of the creation of innovative research outlets for faculty.

To conclude, the emphasis of this section is the need to develop a set of IO training standards that will help attract qualified personnel and students to the profession, as well as the development of a sufficiently large and well-informed faculty to guide education, training, and research programs for these personnel and students. In addition, the authors desire to improve the infrastructure needed to support such programs and strengthen ties between industry, government, and academia through joint education, training, and research initiatives and opportunities.

CHAPTER 7

Information Operations and the Average Citizen

David Wolfe, CISSP, Team Lead, Naval Network Warfare Command, Honeywell

In the space of a few short decades, the explosion in communications and computer technology has turned the Internet into an indispensable information resource for the average citizen. Unfortunately, the very technologies that provide unprecedented access to information and services on the Internet also allow an attacker to electronically enter an individual's home via the personal computer (PC). Once inside, an attacker can gather unseen personal information unknowingly abandoned on the PC's hard drive and use it to wage a highly effective information operations (IO) campaign against even the most careful and secretive computer user.

"Intelligence is the bedrock of IO. It is both foundational and essential to all military operations and its importance to IO is crucial."[1] Arguably, this quote applies equally to business, politics, or any endeavor where an information advantage is sought over others. But influencing an individual, his or her key advisers and confidants, or others who have an impact on the individual's decisions is difficult without relevant, reliable and timely intelligence. To effectively attack an individual requires an in-depth understanding of the targeted individual. Detailed knowledge of the individual's interests, relationships, biases, values, priorities, knowledge and awareness of key facts, and other first-hand information greatly simplifies the process—*if* one can gain direct access to the individual to gather such sensitive information.

In movies and espionage novels, agents employ trusted insiders who have unrestricted access to the individual and the individual's living areas to plant hidden microphones and cameras, allowing the attackers to monitor the individual's activities and conversations in real time. With this unprecedented insider access, the attacker can observe the individual at length to pick up habits and personality quirks. At the same time, the attacker's inside agent dutifully reports exactly what the individual knows (or doesn't know) by identifying which news sources the individual uses and trusts, what documents were read and their source, the names of everyone with whom he or she communicates, and even the content of those private communications. If this level of exploitation seems unattainable (except in the movies), consider this fact: Every bit of information mentioned above, and much more, can be gathered via a home PC in any Western nation with robust Internet access. And getting inside undetected has become an almost trivial task, for some attackers have recently been observed taking over brand new home PCs in as little as *30 seconds* from the time the PC was connected to the Internet![2]

Background

During the early days of the Internet, few ventured into cyberspace because it was expensive, fraught with technical challenges, and offered little to the user beyond the personal satisfaction of being in uncharted territory. Few practical applications existed for home users, so the online community was small, typically technical, and bent on exploring and sharing information about their newfound medium with others. Much like the Ham radio operators of the day, many derived pleasure from making new contacts around the globe, helping other like-minded enthusiasts overcome technical issues, or just enjoying the experience of exploring new frontiers. But unlike the small, well-mannered (and government regulated) Ham community, great leaps in technology during the 1990s brought the Internet into almost every household, but with few limits on content or behavior.

Initially, the focus of home-application developers, service providers, and users was on improving Internet access and creating practical services, not security, because the home computer was little more than an appliance used to reach out to the truly powerful computing and knowledge resources. Aggravatingly slow dial-up access eventually gave way to high-speed digital subscriber lines (DSL) and cable modems, providing virtual-freeway access to the masses for commerce, information, education, and entertainment. The availability of faster processors and modems similarly drove the demand for more robust browsers and e-mail applications to enhance the experience, but security practitioners often lamented in security forums that security was added after the fact, or "brushed on" rather than "baked in" during development. A popular axiom that may explain this lag in security states, "security and functionality exist in an inverse relationship. In other words, the more secure, the less functional, and conversely, the more functional, the less secure."[3] Every added security measure invariably brings with it a cost in performance or additional burden to the user in the form of extra actions, such as entering passwords.

As technology advanced during the late 1990s, the number of practical home web-based applications mushroomed, and enabled the average citizen to directly access and manage virtually every aspect of his or her life with the aid of a PC from the comfort of home. Everything from online banking, stock and commodity trading, shopping, driver's license renewal, and filing taxes to accredited education became possible from anywhere in the world. However, as Frank Abagnale Jr., subject of the Steven Spielberg film, *Catch Me if You Can*, so aptly put it, "technology breeds crime,"[4] and opportunists soon learned to use the Internet to prey upon others at home. Hackers, the earliest abusers of the Internet, discovered ways to use others' new-found high-speed access, storage, and computing power for free, while the criminal element learned to steal information, money, and the very elements of individual identity for use in other crimes. Predatory and manipulative human behavior hasn't changed much over the centuries, just the methods and tools.

The Home PC: An Avenue of Attack or the Target?

Rapidity is the essence of war: take advantage of the enemy's unreadiness, make your way by unexpected routes, and attack unguarded spots.

Sun Tzu[5]

Until recently, most personal computer owners felt relatively safe using new digital technologies and the Internet, for online threats to the home user appeared to be both few and avoidable. Viruses that damaged data and infected other PCs were the first widely recognized threat to the home PC user, but generally could be avoided by refusing to use unidentified floppies or open executable e-mail attachments containing malicious code. Malicious hackers, or crackers as some prefer, were active online, but seldom bothered the home user unless irritated by an action or some actual or perceived insult. The home PC had little to offer a hacker, for it possessed little real computing power and was typically connected to the Internet by a slow dial-up modem, which made it difficult to reach (and exploit) on a predictable basis. That isn't to say home PCs were immune from attack, for hackers used them as convenient hop points as they hopped from compromised PC to compromised PC around the globe to mask their identity and point of origin. This also hindered law enforcement by passing through multiple jurisdictions; a security compromise of an Internet service provider in the U.S. by a hacker may have led to criminal prosecution, but had absolutely no ramifications in another country[6]. But by and large, hacker activity focused on much more powerful university, government, and corporate computer systems and networks, which offered exponentially greater computing power for the recreational hacker, and access to valuable information and databases for the malicious and criminal elements. Databases containing thousands of customers' credit cards or a telephone company's calling card database represented the Holy Grail, so to speak, to the criminal hacker, for a single covertly downloaded database was and remains instantly convertible into cash. These big government and commercial networks and websites also offered a genuine challenge for the recreational hacker, who matched wits with the professional system administrators and security practitioners who invariably defended them.

Throughout the 1990s, the vast majority of hackers seemed primarily motivated by this individual challenge, since success was proof of individual technical superiority, and brought with it bragging rights—and an instant boost to the hacker's ego. Often, hackers defaced high-profile government and corporate websites, then archived and displayed an image of the defaced website online, along with the hacker's screen name, for all to see. To get a feel for the current level of website hacking activity, take a quick trip to the Zone-H website (www.zone-h.org) for a minute-by-minute listing of website hacks around the globe. By 23:55 GMT on 4 September 2005, one of the Zone-H mirror sites posted 1,325 attacks, of which 354 were single IP addresses and 998 were mass defacements *for that day*.[7] Granted, hacking a single server containing individual or personal webpages can yield hundreds or thousands to the count by executing a mass defacement, such as a reported defacement of 306 student webpages on a Stanford University website by Brazilian hackers[8].

Regardless of the motivation, technology has radically altered the hacker paradigm. Hacking is no longer limited to the technical elite, for a plethora of point-and-click tools are available to anyone, tools that require little or no computer knowledge to employ. These easily used, automated tools can be launched from common home PCs that already rival early supercomputers in raw computing power, and grow more powerful each day (Gartner estimates home computers will have 40-GHz processors and 1.3 terabytes of storage by 2008).[9] This combination of advanced, automated hacker tools, powerful home PCs, and "always-on" high bandwidth cable and DSL connections to the Internet gives an attacker unrivaled access and almost limitless avenues of approach to attack any target around the globe. Where pre-Internet generations of attackers and agents gathered intelligence about their intended target through physical reconnaissance and covert surveillance, picked locks, and actually entered the individual's home to install bugs on phones, then unlocked a window or back door to allow easy reentry, the same can now be done from thousands of miles away through a PC. And as mentioned above, much of it is done with the click of a mouse and little or no foreknowledge of the victim.

The Battlefield Moves into the Home

At the end of the last Millennium, governments, industries, and individual computer users breathed a collective sigh of relief when the dire predictions of a Y2K, worldwide computer crash (and the expected opportunistic hacker exploits) failed to materialize. Those of us who sat in command posts that night, linked by multiple, redundant communications paths over networks, satellites, and plain old telephones ran our checklists and followed the New Year as it crept around the globe time zone by time zone, calling command post controllers in each zone just before midnight and listening for the first signs of trouble. By dawn, we knew the extensive (and expensive) measures taken to fix the infamous Y2K bug in our critical system and network software had been effective, and we went home to face the "normal" threats from hackers and virus writers.

But early in the new millennium a shift in attacks, tools, and targets gradually became evident. The rampant growth of annoying and sometimes malicious pop-up advertisement windows on PC screens and the unseen depositing and reading of cookies on users' hard drives were the first hints that strangers could control processes on others' machines and export information without permission. Without realizing it, we were witnessing the birth of spyware, "programming that is put in someone's computer to secretly gather information about the user and relay it to advertisers or other interested parties." [10]

Constantly increasing processor power, coupled with readily available, affordable high-speed Internet access, finally made the home PC a useful target for malicious and criminal attackers. Hijacked by spammers, many "always-on" home PCs were soon pressed into service relaying unwanted advertisements. Even worse, other compromised PCs were turned into mindless zombies or "bots" (short for robots), controlled over Internet relay chat (IRC) or other covert communications channels by distant masters who accumulated laterally thousands of PC soldiers in massive networks of bots, or botnets. This practice has become so common that some security consultants warn that attackers are building botnets comprised of over one hundred thousand bots. [11] With a few keystrokes, these bot armies can be unleashed in a massive coordinated denial of service (DoS) attack against any website, server, or computer on the Internet.

Home PCs still face traditional virus threats, which have evolved to a point where they can be designed for a custom attack using any of a number of drag and drop virus toolkits. [12] To make matters worse, new generations of self-propagating worms appeared on the scene to crawl around the globe, scanning for vulnerable targets and eventually infecting hundreds of thousands of computers in minutes. Code Red, Nimda, Blaster, SQL Slammer, Sasser, Bagle, and other destructive worms were followed by countless variants and new worms. Sadly, the malicious worm and virus writers eventually combined viruses with worms, keystroke loggers, Trojan horses, and other payloads to create a wide range of invasive and/or destructive blended attacks.

Just when home users were beginning to understand the threats presented by worms, viruses, spyware, Trojan horse programs, and bots, the criminal element introduced a new ploy designed specifically to exploit the home user. Exploding onto the Internet in 2004 and growing at an astounding rate, phishing took online trickery to new heights. Phishing is defined by the National Cyber Security Alliance (NCSA) as follows: "Where a perpetrator sends out legitimate-looking e-mails appearing to come from some of the Web's biggest sites, including, but not limited to eBay, PayPal, MSN, Yahoo, BestBuy, and America Online, in an effort to phish (pronounced "fish") for personal and financial information from a recipient." [13]

Avenues of Attack

Avenues of attack are many and varied, changing and evolving daily to take advantage of not just newly discovered technical deficiencies in hardware and software, but devious manipulation of the users themselves. While most home PC users picture their attacker as a stranger halfway across the country or on the other side of the world, the attacker may be someone living in the same neighborhood or parked outside with a laptop computer attempting to gain instant access via wireless networking technology. Armed with just the individual's street address (easily looked up on-line), a laptop computer and a wireless network interface card (NIC), the attacker can simply park near the individual's home and use the laptop and a few readily available software tools to search for a connection over the airwaves. If the targeted home user has installed a wireless network in the home using a wireless router and NICs, and failed to properly configure the security settings, the attacker can connect directly to the individual's home network just as quickly and efficiently as if he walked into the home and plugged an Ethernet cable into the back of the PC. From that instant, the attacker can quickly map and exploit every PC and device connected to the home network at his or her leisure.

For home users who only connect to the Internet via copper wire or cable, the attacker can still conduct reconnaissance and remotely break into the PC from anywhere in the world. Using information harvested from the Internet using popular search engines like Google and various look-up services, the attacker can easily establish the individual's online identity. Once the attacker knows where the individual "lives" online, a quick probe can determine whether the individual is using Window, Unix, Linux, or another operating system, and select the right tools to take advantage of that particular operating system. To compare this with the physical world, this probing is not unlike a burglar who knows that all homes built in a neighborhood either have Kwikset or Schlage locks. He sneaks up to the door at night and tries to insert a Kwikset key. If it slides into the channel smoothly, he knows it is a Kwikset and selects the right tools to pick Kwikset locks. Like the burglar, the online attacker uses stealth and tools crafted specifically to take advantage of known vulnerabilities in the PC's operating system, Internet browser, or other popular applications such as instant messaging.

As mentioned above, avenues of attack are many and varied, giving the attacker an almost unlimited number of ways to probe and attack anonymously. For example, the attacker may choose to launch an automated attack using malicious software tools from another hacked computer and melt back into cyberspace to await the results. One such automated attack method employs computer worms, computer programs that, once launched by the attacker, operate without outside control and propagate complete, functioning copies of themselves onto other PCs to begin the attack cycle again. In the hands of an attacker, worms do more than just replicate; they often carry malicious payloads—sometimes more than one. The payloads may be small programs that collect passwords, install backdoors to allow remote access by the attacker to the PC, or even disable the PC's antivirus software to prevent its own discovery. A recent example of a multi-payload worm that could be used to gain remote access to an individual's PC is one identified by the antivirus firm Sophos as W32.Crowt-A. An astute attacker could send this worm, which masquerades as a news alert from CNN, to everyone associated with the targeted home user. This crafty worm pulls actual current headlines and text from the CNN website to give itself the appearance of legitimacy. When the recipient unwittingly opens the e-mail, Crowt-A acts just like the traditional bug-planting intruder and installs a backdoor (Trojan horse) program, captures the victim's keystrokes, then sends them back to the attacker.[14]And it's invisible to the home user. With a backdoor in place, the attacker can return and enter the compromised PC at will to take total control—and enjoy complete access to any and all information.

For exploits like Crowt-A to be most effective, the attacker should know something about the victim to tailor the e-mail's words to entice the recipient into opening the worm's payload. This act of preying upon human nature is commonly referred to as social engineering[15]. In this case, the attacker exploits the home user's appetite for breaking news to install malicious code on the PC without the user's knowledge. Many of the worms replicating across the Internet are disguised as e-mails from acquaintances to take advantage of the trust between friends, and do this by successfully attacking one PC, harvesting the address/contacts lists from the compromised PC, and e-mailing itself to unsuspecting friends. If the attacker pairs a carefully crafted message with a worm such as W32.Rbot-GR, the attacker not only gains access to the information on the PC's hard drive, but can steal passwords and spy on the user through the PC's own Web camera and microphone.[16]

Although worms are most effective for autonomous, mass-mailing attacks against a large target population, it may be more effective to use a simpler approach when the home user and his circle of acquaintances are already well defined—especially if a close friend's PC is already compromised. Attacking from the friend's PC, the social engineering attacker uses personal information found on that PC to craft a personalized e-mail pretending to be the acquaintance. Since the home user is likely to open the e-mail from someone he knows, the attacker can embed malicious code in the html code of the e-mail itself. In this attack, the simple act of reading the e-mail launches the malicious code and compromises the PC. Another ploy uses personalized information from the friend's PC to entice the home user to click on a link to a website salted with malicious code, which automatically installs itself on the PC's hard drive without the owner's knowledge.

If the attacker's social engineering attempts fail, he may have to resort to other direct attacks that exploit flaws in the PC's operating system, Internet browser, or common application to gain access. Since the most commonly used (and most frequently exploited) operating system and Internet browser are Microsoft Windows (98, ME, 2000, XP) and Internet Explorer, respectively, a multitude of attack tools that specifically target Microsoft products already exist. However, virtually all operating systems, browsers, and software applications contain flaws that can and are exploited, with new ones discovered daily. The attacker can take advantage of the sad fact that many users fail to download and install free security patches for existing flaws, leaving the PC vulnerable to attack by viruses and worms months after the "cure" was released in the form of a patch. For example, the Netsky-P worm, which exploits most versions of Microsoft Windows, was discovered in March 2004. A patch for the flaw was issued and installed by many, yet the malicious software claimed at least two million victims through November 2004 and was still actively probing the Internet and infecting vulnerable PCs.[17]

Other infamous worms, like Mydoom, Sasser, and their many offspring, are similarly still scanning the Internet for PCs vulnerable to specific software flaws. They remain dangerous because their relentless, mindless search for vulnerable PCs continues to find PCs that haven't been fully patched or protected with a firewall before connecting to the Internet. In a recent test that also demonstrates just how quickly PCs can fall prey to older worms written to exploit long-patched vulnerabilities, the firm AvanteGarde connected a half dozen PCs to the Internet, each loaded with a different out-of-the-box operating systems configured with default settings, but without the security patches. This is the configuration a PC would have if the PC owner was reinstalling the operating system and connecting to the Internet to download the patches. Sadly, Windows XP with Service Pack 1 was hacked in as little as 30 seconds, and averaged a 4 minute time-to-hack for the entire test period. However, the newer version, XP with Service Pack 2, which comes with its own embedded firewall turned on by default, wasn't hacked during the two-week test.[18]

The bottom line is that any PC connected directly to the Internet can be exploited by automated attacks under the right conditions, and sooner or later each of us will have the unfortunate experience of being hacked or infiltrated by spyware. The number of new vulnerabilities found every day in the operating systems, browsers, and application software is overwhelming, and the authors of the worms, virus attacks, phishing ploys, and spyware now exploit these vulner-abilities in a matter of *hours*. If this statement seems far-fetched, consider the Windows Plug and Play vulnerability announced by Microsoft on 9 August 2005. Microsoft simultaneously released a patch to fix the vulnerability, but exploit code appeared on 11 August, followed by the Zotob worm and backdoor Trojan on 14 August.[19]

No matter which exploit the attacker uses to get inside the home user's PC, the attacker has the same access to the information stored on the hard drive as the home user—every document, e-mail, and photo—as well as control of every func-tion. Along with that access comes the ability to manipulate everything from the information on the PC's hard drive to deciding what the home user is permitted to see and do on his or her own PC. At this point, the attacker determines reality, for the confidentiality, availability, and integrity of the information on the PC are whatever the hacker decides, and all vestiges of privacy are gone.

What Can an Attacker Really Find on the Hard Drive?

Once the attacker gains access to the home user's hard drive, any document, photo, e-mail, and information found on the hard drive can be remotely downloaded and exploited in moments, and anything typed from this point forward can be collected character-by-character using keystroke loggers. Contacts, e-mail, and e-mail ad-dresses stored on the PC give the attacker a solid foundation for developing an influence model, in which relationships are identified to determine who is able to influence, or conversely is influenced by, the home user. This information can also be used to identify an expanded target set and establish priorities for collection efforts. Passwords are ripe for compromise, for the password files themselves can be downloaded and cracked off-line. As mentioned earlier, the attacker can also turn on Web cams and microphones to gain real-time intelligence about the home user, visitors, and activities within the room, and even the layout and furnishings of the room itself. Clearly, gaining direct, remote access to the home user's PC presents the attacker with tremendous opportunities to collect intelligence about an individual.

But suppose the home user is ultra secretive (some may say paranoid) and saves files almost exclusively to removable media, refusing to store any sensitive files on the PC hard drive. What can the attacker learn if absolutely none of the typical e-mail or office files are saved to the home user's hard drive? Surprisingly, quite a bit, especially if the home user uses virtually any version of the Microsoft's operating system, Internet browser, and office applications. While these popular applications offer many useful features and solid performance, these programs also leave a great deal of unseen or hidden information on the user's hard drive without the user's knowledge, information that can be quickly located and exploited by an attacker who knows where to look. In the following sections we will examine actual examples of unseen information found on PCs running the most commonly used versions of the Microsoft Windows operating system, Microsoft's Internet Explorer browser, and various versions of Microsoft Office. This unseen information is grouped in three broad categories:

1.   Unseen Files. Unseen by the user, Windows and other applications constantly write information to hard drives as a normal by-product of handling files. This includes entire documents saved in their original formats.
2.   Hidden Data Retained Inside Files. Supposedly deleted data is left within files and documents, and is often easily recovered without special utilities.
3.   Clues to Other Information. Convenience "features" in the operating system and browser continuously update lists of recently accessed files and record user actions, providing an electronic trail of breadcrumbs to the home user's sensitive information.

UNSEEN FILES

Temporary (.tmp) Files
In the past, Windows and Windows-based applications earned a reputation for wasting disk space by leaving temporary files on the hard drive after completing tasks. In the early DOS/Windows days when hard drives had limited capacity and a high cost, network administrators conserved precious workstation hard drive space by adding a couple commands at startup to automatically jump to the temp directory, then delete all temporary files. Back then, the focus was on resources, not security.

According to Microsoft, "a temporary file is a file that is created to temporarily store information in order to free memory for other purposes, or to act as a safety net to prevent data loss when a program performs certain functions. For example, Word determines automatically where and when it needs to create temporary files. The temporary files only exist during the current session of Word. When Word is shut down in a normal fashion, all temporary files are first closed and then deleted." [20]

However, when Word or other Windows applications shut down *abnormally* (like when the application/system locks up for no apparent reason), a copy of the working file is often left behind on the hard drive as a temporary file. In general, temporary files result from poorly written programs, improper shutdowns, program hangs, and computer crashes.[21] These temporary files can be used to recover the file being edited at the time of a lock-up, but when users habitually fail to open and close files in the way the application writers envisioned, the result is an ever-growing collection of temporary files scattered across the hard drive. For example, the author used Microsoft Word for Windows to draft a manuscript, working nightly over a period of months without bothering to clean up the temporary files. Not surprisingly, literally dozens of copies of the manuscript were later found, automatically saved in the temp directory with a .tmp file extension or a temporary name beginning with the ~ symbol, and each easily recoverable by an attacker. Finding these temporary files on the hard drive is done by clicking the *Start* button, *Search*, *For Files and Folders*, and typing "*.tmp" or "~*.*". The files will be displayed in the results window, and can be accessed from there, too. Unfortunately, Search doesn't identify temporary files stashed in Temporary Internet Files folders, so this clearly isn't the complete answer!

Complete Copies of Original Files

If the home user regularly downloads files from websites, there is a good possibility he or she took shortcuts or performed other actions that caused the program to write the file to the temp directory. The extent to which this occurs depends on a number of factors, such as the type of operating system used, browser settings, and whether any tools are used to "clean up" after ones self.

To get a sense for the severity of the problem, the author examined the temp directory of a networked Windows 2000 office workstation (viewed with Windows Explorer), and found complete copies of previously downloaded documents in Word and Adobe Acrobat format, all left behind without the author's knowledge and readily exploited by an attacker.

E-mail Attachments in Temporary Internet Files

While the files mentioned above were the result of downloads from websites/portals, complete files are also left behind when working with e-mail. Unseen by the user, copies of files are often written to the hard drive while opening an e-mail attachment. For example, the author e-mailed a draft paper home, then opened it and began editing before saving to another file folder under another title. Consequently, Windows saved copies of both document versions in a folder under Temporary Internet Files. How can the user tell if this happens? During the *Save As* function, the name of the temporary folder (often an unintelligible combinations of letters and numbers) and its contents are revealed and listed on-screen in the *Save As* dialog box.

But even armed with the name of the temporary folder, how does the user know exactly where Windows put these files since they're not left in the root temp folder? The user can do a little on-the-spot investigating simply by clicking on the drop-down arrow of the *Save in* box to display the complete path to the folder—as well as other temporary folders created during previous sessions—then go clean them out. Be advised that looking into the other Temporary Internet Files folders will probably provide a shock, as the author found when he followed his own advice and discovered copies of resumes in other such folders, resumes that were e-mailed to friends months earlier.

Unless the user understands how Windows and other common applications abandon files in unseen locations across the hard drive, sensitive information will accumulate in predictable locations easily exploited by an attacker.

HIDDEN DATA RETAINED INSIDE FILES
Sensitive information is often hiding in plain sight in the very documents we use daily. Because software applications are so powerful and complex with a wealth of functions and features, many home PC users aren't even aware that some features are running in the background, accumulating information that can be exploited by an attacker.

File Properties
Can a single PowerPoint presentation or Word document tell the attacker who provided sensitive information to a home user? Just click on *File*, then *Properties*, and the *Summary* tab will reveal the identity of the person who drafted the original document. This information could lead the attacker to new targets, or the attacker could determine relationships with others just by harvesting names from documents' *Properties* data.

Retaining Deleted/Changed Information
When documents grow inexplicably larger with each save during editing, even after large graphics are replaced with smaller ones, it should be a big hint that something is going on inside the document. Some of this growth may be attributed to the *Undo* function, which records recent changes and deletions on a first-in, first-out basis. Unfortunately, this accumulation of unseen bits and bytes is a normal feature of Microsoft Word and other office applications, which hang onto deleted or replaced data until the *Save As* function is used to create a fresh baseline copy of the file. For example, a home user wants to e-mail portions of a sensitive document to an associate, so he cuts the sensitive information from the document and clicks *File, Send To, Mail Recipient as Attachment*, but fails to first use the *Save As* function. The document visually appears to be sanitized, but the e-mail recipient (and an attacker in the middle) may be able to recover every deleted

detail of the original document, particularly if the home user neglected to turn off the *Track Changes* function under *Tools*. Does it happen? Consider a resume this author recently received from a recent college graduate. After noting a different name in *Properties* and turning to *Tools, Track Changes, Highlight Changes, Highlight Changes on Screen*, the author noted two other individuals used the same resume as a template, replacing the original applicant's information with their own. All three versions were visible (in different colors), providing complete contact information for each of the young ladies—a stalker's delight! Fortunately, this vulnerability is partially addressed in Microsoft Office 2003, which warns the user of the presence of comments and tracked changes during the Save process.

Although an attacker can easily harvest this information from documents, a more detailed examination by a trained professional can yield even more valuable information. In the case of a controversial British government dossier used to justify British participation in the 2003 Iraq War, an IT researcher downloaded a copy of the file in its original Word format from the government website. After writing a small utility to extract the normally inaccessible revision log from the file, he was able to view the user and file names associated with the last ten revisions, revealing the original authors.[22] As the Blair government learned, the only sure way to prevent exploitation is to use an application like Adobe Acrobat Distiller to convert the document into a file format that is basically an image, free of hidden or embedded information.

CLUES TO OTHER INFORMATION

With access to a user's PC, just a few mouse clicks can reveal a tremendous amount of information about the user's activities, and provide the attacker with a roadmap for future collection efforts. And it all comes from existing files without any special hacker tools or advanced technical skills.

Recently Viewed Documents

Right out of the box, Microsoft's word processor, spreadsheet, and presentation applications are tattletales regarding user activities. An attacker need only open the application and click *File* to view a list of all documents recently opened—in order—at the bottom of that drop-down menu. It's a useful tool for users, allowing one to quickly click on the title of the last file or two and resume work. But it also provides the attacker with a list of files the user has been reading and the path to each file's location. Fortunately, it can be easily turned off by clicking *Tools*, then *Options*, select the *General* tab, and uncheck the *Recently used files list*.

But the *Recently used files list* in each application isn't the only place where file use is being tracked by versions of Microsoft Windows and/or Office. Clicking *Start*, then *Documents* gives the attacker a snapshot of all recently opened files of all types, not just Word or PowerPoint documents. Happily, a couple clicks clear this list, too. For Windows 2000 and ME, left-click *Start, Settings, Taskbar and Start Menu*, then click the *Advanced* tab, and click *Clear*. Not only does this clear the list of recently accessed documents, but it also clears the lists of recently accessed websites and programs, too. On the down side, this has to be done every time before shut down to eliminate fresh accumulation.

The Recent Folder

A clean *Documents* list won't slow an attacker with access to your hard drive; the unabridged listing of recently viewed files is stored in a hidden folder in the Windows directory, appropriately named Recent. Not only does the Recent folder give the attacker an interesting list of file names to research, but also lists files on removable disks/USB drives that have been used with the workstation—another valuable hint that important files may be stored elsewhere!

Note: The Recent folder isn't normally visible when using Windows Explorer because Microsoft's default installation prevents users from seeing (and possibly deleting) certain folders (including Recent) and important system files. To view it, click *Tools, Folder Options, View*, then click *Show hidden files and folders.*[23]

The Browser History Folder

Just as the *Documents* tab and the Recent folder show the attacker where to focus, the History folder can act as a roadmap to sensitive information. Using the same analytical skills as a concerned parent checking on a child's online surfing habits, the attacker can use the History folder to track the home user's online activities, determine his or her interests, and even discover personal and professional relationships just by examining which websites the home user accessed.

Eliminating this vulnerability is quick and easy, however. In the Internet Explorer *Tools* menu, click *Internet Options*, and then click the *General* tab. Under *History*, change the number of days *History* tracks the websites visited to zero and click *Clear History* to delete anything already stored.

Browser Favorites

While the History folder reveals when and where the home user has recently been online, the Favorites folder may provide a more complete picture of the home user's actions and interests. Users tend to collect and organize links to websites needed to do their job, as well as links that support personal activities and interests. By analyzing the Favorites folder names and the links within them, the attacker may find clues leading to even bigger pieces of the puzzle. However, a smart home user may not use a favorite when visiting a sensitive website, and can even use a trick to visit the sensitive site in a way that won't be recorded by Internet Explorer. In the browser window, the home user can simply press *Ctrl* plus the letter *O* to bring up the *Open* dialog box, then type in the URL.[24] Now the browser's address drop-down will remain blank and yield no clues.

Cookies

Cookies are yet another valuable source of information for the attacker. Just like the *History* and *Favorites* folders, the accumulated cookies with their embedded URLs give the attacker a comprehensive listing of the home user's on-line activities. How easily can an attacker view the list of cookies? In the browser, click *Tools, Internet Options,* and under the *General* tab click the *Settings* button in the *Temporary Internet files* area. In the *Settings* box, click the *View Files* button.[25] The attacker now has the complete list of cookies and embedded URLs to analyze.

Eliminating every cookie may be appealing, but isn't always practical, because some cookies are needed by some websites to function effectively. However, if privacy is at stake, they must go. To quickly delete existing cookies, click the *Tools* menu in the browser, then *Internet Options*, and select the *General* tab. In the *Temporary Internet files* section, just click *Delete Cookies.* Rather than repeatedly opening the browser to this menu and deleting cookies, it may be easier to block them by clicking the *Privacy* tab and moving the *settings* slider up a notch to block more or all cookies so they don't accumulate in the first place.

Temporary Internet Files

In addition to the documents and cookies mentioned earlier, Temporary Internet Files folders accumulate virtually every object, image, or photo (usually in jpg format) from visited web pages. The objects are automatically saved onto the local hard drive so they won't have to be downloaded again when the user returns to the same website, a time and bandwidth saving measure). An attacker can retrieve the images stored in the Temporary Internet Files and quickly deduce what products the user is researching online, for example.

History Files in Other Applications

To be fair, it isn't just Microsoft products that retain clues to the user's activities. Many Windows-compatible applications retain information, such as Adobe Acrobat, Google Toolbar, Visual Basic, and WinZip, to name just a few. Even basic functionalities such as *Start Menu Run, Start Menu Search*, and *RegEdit* retain information of recent activities.

Remembered Password

Many applications offer to help by "remembering" passwords as they are entered into applications and login screens so users won't have to rely on human memory or keep lists of literally dozens of passwords. Behind the scenes, these passwords are stored in an encrypted file to prevent someone from stumbling across them in their plain text form. If this seems like a safe and easy method, DON'T DO IT! Decline this offer and find another way to safely store your passwords, because a hacker can simply download the encrypted password file and use the same hash algorithm and a password cracking program to generate passwords using every conceivable dictionary word, name, or random combination until he generates one that matches the encrypted value found on the user's password file. Game over.

Cleaning Up and Prevention

Clearly, with all these possible information leaks on the user's PC, remembering to clear these and others—and even remembering *how*—can be a daunting task. Rather than plaguing oneself with opening each application and purging residual information manually, the best answer is to configure the browser, e-mail, and other applications using the steps mentioned above to prevent the buildup of information in the first place. While this provides a good starting point, the real key is defense-in-depth, where home users employ additional automated tools from reputable sources and run them regularly. Fortunately, there are quite a few reputable free/shareware privacy tools already available, and many more anti-virus and anti-spyware products are increasingly being built into most of our favorite utility suites. For example, the Microsoft Anti-Spyware Tool (free Beta version) includes an *Advanced Tools*® tab with a tool called *Tracks Eraser*®, which allows the user to select individual applications and purge their respective history and recent files, or select and purge all with a single click—including the Recycle Bin, which is frequently overlooked. Similarly, Norton System Works includes the *Norton Cleanup*® utility to purge cookies, history, cache, and temporary files. Webroot offers a similar, capable product appropriately named *Window Washer*®. Even non-Microsoft browsers like Firefox and the personal firewall, Zone Alarm Pro now include privacy settings to prevent the buildup of information, as well as cache cleaning tools. Since no single tool can offer complete security for every user and every combination of applications, it's important to explore existing applications for built-in privacy and cleansing functionality, and augment them with a combination of tools to achieve effective defense-in-depth.

Defense-in-depth doesn't stop with finding a mix of tools to purge residual information from our home PCs. Each PC must be defended with a combination of hardware and software tools. A software firewall, coupled with continuously updated anti-virus software, anti-spyware software, and anti-spam software, provide much needed depth to the home users' defensive measures. Why include anti-spam software? Because they carry the bulk of the malicious enticements directly to the users' inboxes, eliminating them before they pique the interest of a household member is essential. Major cable and DSL Internet service providers have begun offering these services as a standard part of their basic service, simplifying the lives of home users who avail themselves of these free offerings. For those who choose to pick their own defensive software, the current generation of anti-virus software and personal security suites offered by the major vendors bundle most of these protections in a single package or suite of products, greatly reducing the need to find individual protective applications that coexist without interference. A full-featured personal firewall that monitors and controls traffic flowing *out* of the PC as well as into it is also a must. After all, how else will you know if an unseen application has been installed on your PC and is attempting to communicate with a stranger?

Wherever practical, the home user should periodically download and run a different tool for each of these essential functions as a double-check. The major anti-virus software vendors all produce effective and reliable anti-spyware products, but sometimes a lesser-known product like Spybot Search and Destroy or AdAware may have different signatures or scan in a manner that exposes unseen threats. And finally, adding an inexpensive router between the home PC and the Internet adds a much needed buffer that prevents unwanted scanning and probing from gathering useful information about your home PC, and also dramatically reduces the amount of traffic that your firewall has to handle. Home-use routers from companies like Netgear, Belkin, and other vendors are relatively easy to install, and don't require someone to program the interfaces or write complicated access control lists. The only thing that *must* be done by the user is to change the default password, for a number of websites offer comprehensive lists of vendor- and equipment-specific default passwords that benefit technicians and hackers alike.

But what happens if all of your defenses fail and you discover a number of strangely named files like Bluesnarf.zip lying around your hard drive? You're lucky, because you probably interrupted the attacker before he had a chance to finish installing his malicious tools and hide his tracks. Unfortunately, the more talented hackers won't leave their tools, or rootkits, in plain sight, so the suspicious home user will have to resort to powerful detective tools to root out the attacker's malware. Or you could just format the hard drive and start over. No matter which solution is selected, regular backups are essential, for sometimes the cure for an infected hard drive can kill your precious data just as surely as a malicious virus. With this in mind, the often overlooked backup is your best and final defense against attack, for it provides the option of cutting one's losses and going back to a known, clean state without the need to spend endless hours trying to remove deeply rooted malware that keeps reinstalling itself no matter what you do. With today's low memory and recordable media prices, backups are no longer difficult, time consuming, or painful to the pocketbook.

Summary

In a perfect world, software would be secure right out of the shrink-wrap and leave the hard drive clear of any unseen information. Fortunately (for the attacker), the most popular operating systems, office applications, and Internet browsers have yet to adequately address privacy concerns within the applications themselves, leaving not just a trail of breadcrumbs for the attacker to follow, but often the crown jewels of information themselves in plain sight. Virtually every Windows-based application used in the modern office and home environment writes information onto the PC's hard drive, often without the user's knowledge, and fails to remove it without operator intervention. This isn't meant as an indictment of the operating system and application developers, for most were developed in a more open, less threatening environment than today's Internet, where attackers are actively seeking personal information for criminal and/or malicious purposes. But until developers design security into the code from day one, service providers take aggressive action against malicious users, and home users implement defense-in-depth measures, attackers will continue to enjoy virtually unfettered access to home PCs to conduct on-line reconnaissance and remote exploitation of the information found there.

Security analysts agree that it isn't a question of *if* the attacker can gain access to the home user's PC and information, but *when*. To quote Consumer Reports, "Use the Internet at home and you have a one-in-three chance of suffering computer damage."[26] In the same article, they pointed out that 66 percent of Windows users surveyed reported detecting a virus in the past two years.[27] The spyware threat is harder to gauge, since it comes in many guises, and often operates quietly in the background without the user's knowledge. Home users in the Consumer Reports survey found spyware on their machines 52 percent of the time,[28] while the National Cyber Alliance reports that spyware affects 80-90 percent of desktops.[29] With these kinds of odds, the advantage is clearly with the attacker.

Once the attacker is inside without the home user's knowledge, he or she is free to conduct reconnaissance and harvest information (including a great deal that is left behind by the current generation of software), and monitor the home user's actions and environment in near-real time. Theft of any and all financial information, passwords, and identity information must be assumed. Armed with this plunder and continued on-line access directly to the home user, an astute attacker can alter reality as seen by the home user by adding, deleting, modifying, or even controlling the flow of information. If the home user is heavily dependant upon online services such as e-mail and Internet-based applications, it goes without saying the attacker can subtly manipulate information and the home user's online experiences to figuratively manipulate the home user like a puppet on strings.

Before becoming completely paranoid and forsaking the home PC forever, consider the home PC in the same light as an automobile. Each model is more technologically advanced than the last, with new features and added safety. Used by an operator with a basic grounding in the law, rules of etiquette, and safety, the odds of a catastrophic event are remote if the user drives defensively and properly maintains the vehicle. The same is true of the home PC user, who can avoid the loss of personal information and equipment damage by adopting a defense-in-depth strategy. Installing and continuously updating protective software, configuring existing applications' built-in options to reduce or eliminate the buildup of unwanted information, and looking at every e-mail, instant message, and website with a critical (cynical) eye toward the attacker's subtle trickery and social engineering ploys will fend off the bulk of the attacks. Sadly, technology will continue to serve up a steady stream of new vulnerabilities, and attackers will dream up even more creative ploys and attack vectors. But if the home user understands the threat and prevents critical personal information from accumulating on the PC hard drive, the attacker gets little or nothing, and subsequent patches and antivirus updates will soon find and eliminate the threat. In the end, it's impossible to defend against every conceivable threat or attack, but it is possible to minimize exposure and risk—and recover.

A Tale of Two Cities: Approaches to Counterterrorism and Critical Infrastructure Protection in Washington, D.C. and Canberra

Jeff Malone, Noetic Solutions Pty. Ltd., Leigh Armistead, Edith Cowan University

O ver the last few years, there has been much written on the potential threat posed by cyberterrorism, the targeting of computer networks and related infrastructures by individuals or groups for terrorist purposes. But much of this literature has been sensationalist, has focused narrowly on technical computer security issues, and has failed to link the discussion of cyberterrorism with the broader issues relating to either terrorism or policy responses to it.[1] Yet is it precisely because of the interdependence between the changing nature of global terrorism and the increasing vulnerability of critical national infrastructure (CNI) that cyberterrorism becomes a plausible threat.[2] This chapter examines the cyberterrorism (CT) and critical infrastructure protection (CIP) policies of the United States and Australia, both before and after 11 September 2001, in light of the changing nature of global terrorism. The paper concludes by offering some explanations for the differences in approaches to CT and CIP policies adopted by each state.

U.S. Counterterrorism and Critical Infrastructure Protection Policies Prior to 11 September

During the cold war, United States national security policy was focused on minimizing the possibility of strategic nuclear attack by the Soviet Union. There was a general understanding of the nature of the threat posed by the Soviet Union, and most of the international security efforts of the United States (and the West in general) were directed at countering it. But with the collapse of the Soviet Union in 1991, and with it the bipolar world order, the strategic certainty provided by this structured threat disappeared. The specter of global nuclear war was replaced by a range of diffuse and unstructured threats and challenges.[3] The reality of the new security environment was brought home to the United States with the bombing of the World Trade Center in February 1993. A little over two years later, the scene was replayed when domestic terrorism struck at the nation's heartland on the morning of 19 April 1995, at the Alfred P. Murrah Federal Building in Oklahoma City.

These events raised awareness of the threat posed by terrorism to the United States, but tangible policy outcomes took a little longer to emerge. The first key Clinton administration response to the evolving terrorist threat was to promulgate Presidential Decision Directive (PDD) 39 *U.S. Policy on Counterterrorism*, dated 21 June 1995.[4] PDD 39 articulated a four-point strategy that sought to reduce vulnerability to terrorist acts, deter terrorism, respond to terrorist acts when they occur, and deny terrorists access to weapons of mass destruction (WMD). The four-point strategy integrated both domestic and international measures to combat terrorism. PDD 39 was novel in that it specifically identified the vulnerability of CNI and potential terrorist employment of WMD as issues for concern.[5] But in general, PDD 39 generally lacked sufficient bureaucratic teeth to achieve meaningful outcomes. Yet a clear outcome of PDD 39 was to raise the profile of CIP policy. U.S. CIP policy was not a novelty per se. But previously CIP policy had tended to be overshadowed by other elements of U.S. national security policy.[6] And the increasing interconnectedness of CNI had created a range of dependencies and vulnerabilities that were historically unprecedented.[7] PDD 39 directed the Attorney General to establish a committee to review and report upon the vulnerability of CNI to terrorism.[8] In turn, this committee—the Critical Infrastructure Working Group (CIWG)—noted that CNI was vulnerable not only to attack by physical means, but conceivably also by computer-based means.[9] The lack of knowledge concerning cyber threats prompted the CIWG to recommend that a presidential commission be established to more fully investigate this matter.

The CIWG's recommendation led to the establishment of the Presidential Commission on Critical Infrastructure Protection (PCCIP) on 15 July 1996, by Executive Order (EO) 13010.[10] While the PCCIP was primarily a response to PDD 39, in an informal sense it also consolidated a range of uncoordinated CIP policy development activities occurring across government sectors.[11] EO 13010 also directed the establishment of an interim Infrastructure Protection Task Force (IPTF) within the Department of Justice (DoJ), chaired by the Federal Bureau of Investigation (FBI).[12] The purpose of the IPTF was to facilitate coordination of existing CIP efforts while the PCCIP undertook its work.[13] The IPTF was chaired by the FBI so that it could draw upon the resources of the Computer Investigations and Infrastructure Threat Assessment Center (CITAC), which had been established within the FBI in 1996.[14] In effect, the IPTF represented the first clear effort to establish coordination across different government agencies and with the private sector for CIP.

The final report of the PCCIP, *Critical Foundations* was released in October 1997.[15] The key finding of the PCCIP was that while there was no immediate overwhelming threat to CNI there was need for action, particularly with respect to national information infrastructure (NII) protection. The PCCIP recommended creation of a national CIP plan, clarification of legal and regulatory issues that might arise out of such a plan, and a greater overall level of public-private cooperation for CIP. From late 1997 to early 1998, the PCCIP report underwent interagency review to determine the administration's policy response. But in February 1998 concrete outcomes were already beginning to emerge, as the interim IPTF was amalgamated with the CITAC, made permanent, and renamed the National Infrastructure Protection Center (NIPC).[16]

The recommendations of the PCCIP were given practical expression on 22 May 1998 with the release of two policy documents: PDD 62 *Combating Terrorism* and PDD 63 *Critical Infrastructure Protection*.[17] These two documents were the culmination of the Clinton administration's efforts at policy development for CT and CIP. PDD 62 was a direct successor to PDD 39. It provided a more defined structure for CT operations, and presented a focused effort to weave the core competencies of several agencies into a comprehensive program.[18] Also in common with PDD 39, PDD 62 sought to integrate the domestic and international elements of U.S. CT policy into a coherent whole.

PDD 62 also established the Office of the National Coordinator for Security, Infrastructure Protection, and Counterterrorism.[19] It was intended that the office would facilitate interagency activities, and in order to achieve this objective it established three senior executive working groups: the CT Security Group, the Critical Infrastructure Coordination Group, and the WMD Preparedness Group. Collectively, the purpose of these working groups was to enhance capability for CT response. But the office did not have the authority to mandate procedures to executive agencies, so its ability to affect change was limited.

PDD 63 was the document that implemented the recommendations of the PCCIP report, as interpreted through the prism of the interagency review panel.[20] PDD 63 identified twelve sectors of CNI, appointed government lead agencies for each of these sectors, and established coordination mechanisms for the implementation of CIP measures across the public-private divide. In particular, PDD 63 vested principle responsibility for the coordinating activities in the Office of the National Coordinator (which had been set up under PDD 62). PDD 63 also established the high level National Infrastructure Assurance Council (NIAC), to advise the president on enhancing the public-private partnership for CIP. PDD 63 also called for a National Infrastructure Assurance Plan (NAIP), which would mesh together individual sector plans into a national framework. Finally, PDD 63 authorized increased resources for the NIPC, and approved the establishment of Information Sharing and Analysis Centers (ISACs) to act as partners to the NIPC.

In the last year of the Clinton administration, there were some minor changes to CT and CIP policies. Version 1.0 of a *National Plan for Information Systems Protection* was released in January 2000.[21] This was the direct result of the call in PDD 63 for a NAIP.[22] But reflecting the priority given to cyber security issues by the PCCIP, it primarily addressed NII protection rather than CIP as whole.[23] In June 2000, the *Terrorism Preparedness Act* established the Office of Terrorism

Preparedness (OTP) within the Executive Office of the President. Its role was to coordinate CT training and response programs across federal agencies and departments. But like the Office of the National Coordinator established by PDD 62, it had to rely on suasion rather than formal authority to achieve its objectives.

When the George W. Bush administration came to power in early 2001, existing CT and CIP arrangements were consolidated. The collection of senior CT and CIP groups became the Counterterrorism and National Preparedness Policy Coordination Committee (PCC), reporting to the National Security Council (NSC).[24] And while some debate occurred on future directions for CT and CIP policy, it bore no fruit prior to 11 September 2001.[25] So in practice, during the first nine months of the George W. Bush administration, the bulk of the CT and CIP arrangements in place in the United States were largely a legacy of the previous Clinton administration.

Across the period leading up to the 11 September attacks, the international aspect of the terrorist threat against the United States was becoming more evident. Incidents that demonstrated the international character of the terrorist threat included the 1993 World Trade Center bombing, the June 1996 attack on the Khobar Towers complex in Saudi Arabia, plans to attack U.S. airliners in Southeast Asia in 1996, the attacks on U.S. embassies in Kenya and Tanzania, and the attack on the USS *Cole* in October 2000. In response to these incidents, both PDD 39 and PDD 62 incorporated measures to combat terrorism abroad. However, the domestic focus of U.S. CT and CIP policies largely overshadowed the international dimension of the evolving terrorist threat.

Australian Counterterrorism and Critical Infrastructure Protection Policies Prior to 11 September

Compared with the United States, the cold war did not loom as large in Australian national security planning. Excepting the extreme possibility of Soviet strategic nuclear attack on Australian soil, Australian security planners largely focused on a variety of lower-level potential threats to Australia in the wake of the Vietnam War. By the mid-1970s, these came to be centered on the archipelagic region to Australia's immediate north. This did not reflect concern that Australia might come into conflict with states in this region. Rather, it recognized the geostrategic reality that any military threat to the Australian continent must come from the north. So this approach to security planning focused on the capabilities required to mount an attack on Australia, rather than a specific threat from an identified adversary.[26] During the same period, Australia had been almost entirely free of domestic politically motivated violence. So there was little sense of threat to Australia from either internal or external sources.

Perceptions of the threat to Australia posed by terrorism changed in February 1978, with the detonation of a bomb outside the Sydney Hilton Hotel at the time of the Commonwealth Heads of Government Meeting (CHOGM). While in global terms the bombing was minor (three dead and eight injured), the incident marked an end of innocence for Australia.[27] In the wake of the incident, the government commissioned a review of Australia's protective security arrangements, including response arrangements for terrorist incidents.[28] The review made a wide variety of recommendations regarding commonwealth-state cooperation, mechanisms for employing the Australian Defense Force (ADF) in support of civil authorities, and enhanced capabilities within the ADF and state police forces for resolving terrorist incidents and responding to incidents involving explosive devices. The review also noted the need to identify and protect CNI as an integral part of CT policy.[29]

The bulk of Justice Hope's recommendations were accepted by government, and came to be expressed principally in the form of the Standing Advisory Committee on Commonwealth/State Cooperation for Protection Against Violence or SAC-PAV.[30] SAC-PAV was given responsibility for the National Anti-Terrorist Plan (NATP), and provided the umbrella under which CT policy, capability development and acquisition, and cooperative and crisis management arrangements might evolve, at both the commonwealth and state levels.[31] Paralleling SAC-PAV, the government also established the Special Inter-Departmental Committee for Protection Against Violence (SIDC-PAV) to coordinate CT policy issues at the commonwealth level. In terms of CIP policy, the key response was the Vital Installations program, which put in place a variety of measures to protect CNI.[32] There were reviews of the SAC-PAV arrangements in 1986 and 1992, but recommendations for changes to these arrangements were minor.[33] So the CT mechanisms that were put in place in the early 1980s remained essentially unchanged until late 2001.

But during the 1990s there was some activity on the CIP policy front. In late 1996, the Defense Signals Directorate (DSD) reported to government on the vulnerability of the Australian NII to a variety of threats and hazards, and recommended further government action.[34] This led to the creation of an Interdepartmental Committee (IDC) that was directed to develop policy for NII protection.[35] The National Security Committee of Cabinet (NSCC) considered the August 1999 IDC report. This culminated in a government announcement of a five-point strategy for NII protection later that month.[36] While these outcomes represented a step forward, concerns were expressed that the pace of government activity was too slow, and the resources devoted to implementing strategy were inadequate.[37]

In the lead-up to the Sydney 2000 Olympic Games, there was some enhancement of intelligence sharing and the procedures for the employment of the ADF (in particular, its reserve component) in support of CT and security operations.[38] There was also recognition of the evolving nature of the terrorist threat, in the raising of the Joint Incident Response Unit (JIRU) within the ADF for response to terrorist employment of WMD.[39] Over the same period, there were some changes to the measures in place for CIP.[40] But these initiatives were effectively minor adjustments to existing policies. And while Australian CT capabilities were improved during the Sydney 2000 Olympic Games, the bulk of this additional capability was disestablished at the Games' completion. So these enhancements were essentially tactical reactions to what was perceived as the specific (and temporary) challenge posed by the Olympic Games, rather than a strategic response to the changing nature of global terrorism.

During the period leading up to 11 September 2001, the international dimension of the evolving terrorist threat was becoming more evident to Australian policymakers. Investigations in the mid-1990s indicated that the Aum Supreme Truth sect (sponsors of the March 1995 Tokyo subway attack employing sarin gas) had tested sarin gas on a remote sheep station in Western Australia.[41] And the Sydney 2000 Olympic Games raised the profile of Australia as a potential terrorist target. But overall, the bulk of Australia's activities to combat terrorism in the international arena at this time consisted of low-key diplomatic activities

U.S. Counterterrorism and Critical Infrastructure Protection Policies after 11 September

The terrorist attacks on 11 September led to fundamental changes to the U.S. government's approach to CT and CIP issues. On 8 October 2001, EO 13228 established the Office of Homeland Security (OHS), to be headed by the adviser to the president for Homeland Security (Tom Ridge, previously the governor of New Jersey).[42] The purpose of the OHS was to develop and coordinate a national strategy to protect the United States against a terrorist attack, in light of the new threat posed by global terrorism.[43] EO 13228 also established a high level Homeland Security Council (HSC) that is responsible for advising the president on all aspects of homeland security.[44]

On 9 October 2001, Gen. Wayne Downing was appointed National Director for Combating Terrorism and Richard Clarke was named Special Adviser to the President for Cyberspace Security.[45] Significantly, Downing had previously been the Commander-in-Chief of the U.S. Special Operations Command (USSOCOM), so his appointment reflected a greater prominence for the international (and overtly military) dimension of U.S. CT policy. Clarke's duties were formally spelled out on 16 October 2001, with the release of EO 13231, which established the President's Critical Infrastructure Protection Board (PCIPB). The PCIPB was to recommend policies and strategies for the protection of critical information systems. The same EO also established the high level National Infrastructure Advisory Council (NIAC) to provide advice to the president on the same matter.[46]

In July 2002, the OHS released the *National Strategy for Homeland Security*.[47] The purpose of the strategy was to integrate all government efforts for the protection of the nation against terrorist attacks of all kinds.[48] In effect, the strategy updated the measures enacted under PDD 63 in light of the post–11 September environment. The strategy did not create any new organizations, but assumed that a Department of Homeland Security (DHS) would be established in the near future.[49] In September 2002, the PCIPB released for comment the draft *National Strategy to Secure Cyberspace*.[50] In effect, this document was the proposed successor to the Clinton administration *National Plan for Information Systems Protection*. The subject of the draft plan was welcomed, but concerns were expressed that it lacked the regulatory teeth to prompt action by the private sector.[51]

The most obvious consequence of the United States' revised approach to CT and CIP occurred in November 2002, with the creation of the DHS.[52] This consolidated the bulk of U.S. federal government agencies dealing with homeland security (consisting of more than 170,000 employees) into one department headed by a cabinet-level official.[53] This represented the most fundamental change to U.S. national security policies since their inception in 1947. The DHS was initially comprised of five directorates (Management, Science and Technology, Information Analysis and Infrastructure Protection, Border and Transportation Security, and Emergency Response and Preparedness).[54] Significantly, the DHS reflected some of the measures that had been proposed by the U.S. Commission on National Security/Twenty-First Century (the Hart–Rudman Commission).[55] But it was only after the events of 11 September that the political imperative for significant organizational change for CT and CIP emerged.

Further action continued into 2003, with the release of three policy documents: the final version of the *National Strategy to Secure Cyberspace*, the *National Strategy for the Physical Protection of Critical Infrastructures and Key Assets*, and the *National Strategy for Combating Terrorism*.[56] At the same time, EO 13286 abolished the PCIPB and the position of Special Adviser on Cyberspace Security.[57] The NIAC was retained, but now reported to the president via the DHS. Combined with the departure of key staff associated with cyber security issues, these measures raised concerns that cyber security issues were being marginalized in the new arrangements. And in the wake of Hurricane Katrina, other concerns were raised that the capacity to respond to threats to CNI from natural hazards had actually decreased since the creation of the DHS.[58] But overall, it remains difficult to assess how effective the measures put in place since 11 September 2001, will be in the event of a significant attack on U.S. CNI, particularly via cyber means. While DHS conducted a major exercise on precisely this scenario (dubbed "Cyber Storm") in February 2006, the lessons to be learned from this exercise far from clear.[59]

Overall, arguably the most evident aspect of the new approach to CT policy has been the involvement of the United States in substantial military campaigns in Afghanistan and Iraq as part of the Global War on Terror (GWOT).[60] But these two military campaigns have tended to overshadow a range of lower-key military and diplomatic activities across the globe.[61] Also, the GWOT does not represent a fundamental departure from the international aspects of CT policy articulated previously in PDD 39 and PDD 62, both of which incorporated a variety of measures to combat terrorism overseas. Rather, the difference between the GWOT and these previous policies is one of emphasis: the changed nature of global terrorism mandated a CT policy with a more evident international dimension.

Australian Counterterrorism and Critical Infrastructure Protection Policies after 11 September

The terrorist attacks on 11 September had an immediate effect on U.S. CT and CIP policies, and the same can be said of Australian policies. The Australian Prime Minister was present in Washington, D.C. on the day of the attack. The imperative for improved CT mechanisms was tragically reinforced a little more than a year later by the 12 October bombing in Bali. While the enhancement of Australia's CT and CIP arrangements was well underway by October 2002, this event provided added impetus for increasing Australia's capacity to respond to the threat posed by global terrorism, particularly in response to attacks against Australian interests and citizens overseas.

An initial review of Australia's CT capabilities in October 2001 made four broad recommendations to cabinet.[62] First, the review recommended an immediate increase in resources for operational agencies, which were now faced with increased demands for their services.[63] Second, the review recommended some organizational realignment (though largely within established departmental boundaries) of a number of CT agencies.[64] Third, the review recommended some long-term enhancement of detection capabilities (particularly baggage handling in the aviation industry) and response capabilities within the ADF.[65] Finally, the review recommended strengthening of Australia's legal regime for dealing with terrorism.[66] By the end of 2002, the federal government had enacted the bulk of these recommendations, and there has continued to be ongoing reform and adjustment of these measures up to the time of writing.

There were also changes to CIP policies. These changes included a recommendation of increased resources and a broadened focus on CNI as a whole (rather than concentrating on NII protection).[67] In particular, new arrangements were made to facilitate cooperation and information sharing between the public and private sectors in support of CIP in the form of a set of agreements collectively referred to as the Trusted Information Sharing Network (TISN).[68] Significantly, many of the adjustments to Australian CIP policies after 11 September 2001 were similar to the measures enacted by PDD 63, as well as the recommendations of the IDC on NII protection dating from 1998. This suggested that while the need to improve Australia's CIP policies had been understood for some time, only now did the political imperative for funding them exist. But criticism continued regarding both the pace and level of funding for CIP activities.[69]

Complementing these reforms (which focused on operational agencies), changes were also made to the high-level coordinating mechanisms for CT.[70] In April 2002, the commonwealth and states agreed to a modification of the SAC-PAV.[71] In the new organization, now called the National Counterterrorism Committee (NCTC), much of the work previously undertaken by SAC-PAV would continue unchanged. But the new arrangements recognized an increased role for the commonwealth, and broadened the mandate of the NCTC to include the prevention of terrorism and consequence management of WMD incidents.[72] Reforms also increased the role of the NCTC with respect to incidents involving CNI.[73]

Around the same time, the SIDC-PAV was split into two new committees: the Commonwealth Counterterrorism Committee (CCTC) and the Commonwealth Counterterrorism Policy Committee (CCTPC). The CCTC was to focus on coordinating the commonwealth's operational response to terrorist incidents. In contrast, the purpose of the CCTPC was to coordinate the development of longer-term CT policy across the commonwealth and oversee the international dimensions of CT policy. Capping the enhancement of coordinating mechanisms, the government established a new National Security Division (NSD) within the Department of Prime Minister and Cabinet (DPMC) in May 2003.[74] One of the key tasks of the NSD is to coordinate CT and CIP policy across government, which it will do by providing the support secretariat for the NCTC and the CCTPC (with the PSCC providing the secretariat for the CCTC).[75]

There were also adjustments in policy directed at the international level. In the wake of the 11 September 2001 attacks, the Australian government invoked the ANZUS Treaty (for the first time in the treaty's history) as the basis for Australia's involvement in the GWOT.[76] This action reflected the recognition that the changed nature of global terrorism necessitated action at the source, wherever this might be: at home, in Australia's region, or further abroad.[77] This was formally articulated as official policy in documents released in 2003 by the Department of Foreign Affairs and Trade (DFAT) [83] and the Department of Defense (DoD).[78] Therefore, one of the key outcomes in Australian CT policy after 11 September 2001 and the Bali bombing was the increased importance placed on the international dimension of Australia's CT efforts.

Australia's highest profile international CT effort was the deployment of elements of the ADF to Afghanistan (Operation Slipper) as part of the GWOT. But there were many other, lower-profile activities occurring at the time. Since late 2001, Australia had been involved in a vigorous campaign to engage its neighbors on both a bilateral and multilateral basis on a range issues relating to combating terrorism. Most notably, this led to the creation of the post of Ambassador for Counterterrorism in March 2003 to provide a focal point for Australia's diplomatic efforts to counter terrorism.[79] The outcomes of these activities have included intelligence-sharing arrangements, mechanisms to counter money-laundering and terrorist financing, and enhancing the skills of regional police and security forces.[80] As with the U.S., these efforts do not represent a fundamental change in Australia's foreign policy regarding terrorism. Rather, they represent a change in profile and priority that has arisen out of the recognition of the increased threat posed by global terrorism in the contemporary era.

Counterterrorist and Critical Infrastructure Protection Policies: The View from Two Cities

The overall thrust of CT and CIP policy in the United States and Australia was essentially similar prior to 11 September 2001.The governments of both states understood that the nature of the threat posed by terrorism was changing. This was manifest by the concerns about potential terrorist employment of WMD, and the apparent vulnerability of CNI to terrorist acts. In both states, CNI policy tended to focus primarily upon the cyber threat to the NII, as far less was known about the contours of this threat than of more traditional terrorist tactics, or even of potential terrorist employment of WMD. And in both states, the focus of operational CT policy tended to be reactive, in that it was focused on responding to terrorist acts rather than shaping the environment that gave birth to terrorism.

Similarities in policy fundamentals continued after 11 September 2001. In the wake of the attack, the resources that both states devoted to CT and CIP activities were markedly increased (though not to the same extent). In both countries, CIP policy expanded from a fairly narrow focus on protecting the NII, to one that encompassed all CNI. And in both the U.S. and Australia, where CT policy had previously focused on domestic response, there was a marked increase in the international scope of CT policy. Most spectacularly, this was manifest in the conduct of major military operations in Afghanistan and (controversially) in Iraq. But in both states, the international aspect of CT policy also involved a range of lower-profile activities to combat terrorism by shaping the environment from which global terrorism sprang.

But despite the similarities in the CT and CIP policies adopted by the United States and Australia, there were notable differences. While these differences preceded the 11 September attack, they became more pronounced after that date. In effect, these differences manifested in three key themes. First, the pace, priority, and relative measure of resources devoted to CT and CIP activities in the United States has been consistently higher than in Australia. Second, whereas CT and CIP policy activities were clearly and publicly announced in the United States, this has not always been the case in Australia. And third, while the demands of enhanced CT and CIP policies led to the most significant overhaul of the U.S. national security machinery since its inception, the organizational response in Australia involved little more than tinkering with existing structures.

There would appear to be three interrelated factors that explain these differences. The first factor is the difference in the level of perceived threat and vulnerability between the United States and Australia. In the United States, the series of terrorist incidents culminating with the 11 September attack created a sense of insecurity unparalleled since the height of the cold war. At the same time, a number of executive-level exercises and incidents heightened official concerns from the mid-1990s onward regarding the vulnerability of CNI. By contrast, terrorist incidents in Australia after the bombing during the 1978 Commonwealth Heads of Government Meeting attracted comparatively little attention. The attacks on 11 September and the Bali Bombing, while shocking, lacked the visceral impact of a major terrorist incident on Australian soil. And while there was growing official concern over the vulnerability of CNI, this was of a much lower order than in the United States. The political imperative for substantial CT and CIP policy action in Australia was, and is, significantly less than in the United States.

The second explanatory factor is the difference in political culture between the two states and their conduct of national security affairs. In the United States there is a strong separation of powers and the conduct of government is relatively open. This means that highly formalized declared policy and organizational actions are essential (and expected) conditions for achieving effective policy outcomes. By contrast, in the Australian political system the executive arm is far more dominant, the conduct of government is significantly less open, and there is a tradition of relative procedural informality. This means that formally articulated policy documents are neither necessarily expected nor required for policy implementation. So the relative absence of formally announced policies, and the ad-hoc organizational reforms and adjustments undertaken to enhance Australia's CT and CIP capabilities reflect the behavioral norms of the Australian political system.

The third factor that impacts Australian and U.S. policy is the magnitude of interagency coordination in each country. As a global superpower and large state, the scale, scope, and complexity of the interagency coordination problem for the United States is significantly greater than that for Australia. Accordingly, in the United States there was strong support for the creation of the Department of Homeland Security, whereas such a measure was considered necessary in Australia. Rather, the government regarded the existing interagency mechanisms for coordinating CT and CIP activities as appropriate in the post–11 September 2001 environment.

Conclusion

Since the early 1990s, the nature of the threat posed by global terrorism has fundamentally changed. And in response to the changing contours of this threat, states across the globe have adjusted their CT and CIP policies. This paper has surveyed the policy directions followed by the United States and Australia. In particular, the paper has noted that in both states the overall profile of CT and CIP policies increased after 11 September 2001, but the level of concern specifically focused on cyber threats declined. While cyberterrorism is presently not regarded as a credible threat, it does not follow that this will always be the case.[81] So if and when a genuine act of cyberterrorism occurs (either alone or in conjunction with a physical attack), it will be interesting to observe how effective the CT and CIP measures put in place in response to the physical attacks of 11 September 2001, will be at coping with this new threat.

Speaking Out of Both Sides of Your Mouth: Approaches to Perception Management in Washington, D.C. and Canberra

Jeff Malone, Noetic Solutions Pty. Ltd.
Leigh Armistead, Edith Cowan University

One of the key implications of the Information Age for global politics is the dramatically increased significance of the broadcasting or projection of information, through a variety of means, as an instrument of power. A variety of terms can be used to describe these activities. Strategic communications, public diplomacy, and perception management (the term that, in the interests of consistency, will henceforth be used in this chapter) are but a few examples of such terms. Irrespective of which term is preferred, they all describe (from a state's perspective) fundamentally the same activities: the projection or dissemination of information to foreign audiences, in order to elicit behaviors in those audiences that support the achievement of national security objectives.

In this chapter, the respective approaches to perception management of the United States and Australia will be examined, with the principal focus being on the organizational aspects of perception management policies. It commences with a brief discussion perception management activities have become so important in the contemporary global security environment, and then examines the approaches to perception management of the United States and Australia, both before and after 11 September 2001. The chapter then compares the approaches to perception management of the United States and Australia, with a view toward offering explanations for similarities and differences between the respective approaches of the two states and then concludes with some remarks on the future prospects for perception management in the context of the ongoing global struggle against transnational terrorism that is commonly referred to as the Global War on Terror (GWOT).

Perception Management in the Information Age

The employment of perception management as an instrument to achieve national security objectives is arguably as old as the history of statecraft itself.[1] So why is it the case that perception management has become so important in the contemporary global security environment? In simple terms, the increased significance of perception management is a direct consequence of the information revolution for the global security environment. Contemporary information and communications technologies allow for the production, transmission, and storage of informational content in ways that are practically impossible for states to control. National boundaries are entirely porous to all sorts of information flows. Barriers (be they natural or artificial) that once shielded domestic audiences from undesired foreign messages simply no longer work.

Complementing the above trend is the democratizing aspect of contemporary information and communications technology, in the sense that it can be acquired and exploited by a wide range of global actors well beyond established states.[2] The reach and access provided by information and communications technology means that perception management provides a cheap and effective, and therefore highly attractive, instrument of power for non-state actors. Similarly, information and communication technology enable the creation and mobilization of virtual organizations and communities—including transnational terrorist groups—across national boundaries.[3] From a state perspective these trends mean that the conduct of perception management activities in support of national security objectives is now both more important, and more challenging. So it is in terms of these trends that the perception management practices of the United States and Australia need to be understood.

Historical Background of Perception Management Arrangements in the United States

The origins of the United States perception management practices date back to World War II, and the Office of War Information (OWI), which was established by executive order in June 1942.[4] The charter of the OWI gave it the mandate to disseminate information (via publications and radio broadcasts) to both domestic and foreign audiences.[5] Objections to the possible propagandizing of domestic audiences—even during a time of total war—led the OWI to focus its activities squarely on foreign target audiences in enemy or neutral states.[6] Clearly, concerns about the targeting (deliberate or incidental) of domestic audiences by national perception management activities were present at the very inception of formal U.S. perception management policies. In common with other agencies established to oversee the conduct of the war, OWI was formally disestablished a few months after the end of hostilities. But some elements formerly contained within OWI would ultimately form part of the perception management program established for the waging of the cold war.[7]

In response to the challenge posed by the Soviet Union, the Truman administration wanted to strengthen and coordinate foreign perception management measures in order to attain U. S. national objectives. To do this, the National Security Council (NSC) passed executive directive NSC 4, *Coordination of Foreign Information Measures* on 17 December 1947.[8] This policy was expressly designed to combat an extensive propaganda campaign being conducted by the Soviet Union at that time. In particular, NSC 4 was written to exploit and promote the message of economic aid that the United States was delivering to a number of foreign nations, especially in Europe.

As the result of the disbandment of the OWI in back in 1945, there was no existing government agency tasked to conduct perception management. Therefore this policy document was meant to serve as the basis for interagency coordination. The plans expressed in *Coordination of Foreign Information Measures* were mainly seen as collateral duty for the newly created Assistant Secretary of State for Public Affairs position, whose other official duties included efforts to counter the effects of anti-U. S. propaganda campaigns domestically.

At the same time, concerns regarding the possibility of propagandizing domestic audiences by perception management activities led Congress to pass the Smith-Mundt Act of 1948, which forbade the State Department from disseminating propaganda to citizens of the United States.[9] The prohibitions put in place by the Smith-Mundt Act remain in place to this day, and therefore still constitute the legal basis for the conduct of perception management activities by the U. S. foreign policy bureaucracy.[10]

This new act created a serious dilemma for the State Department. The Assistant Secretary, who is supposed to conduct public diplomacy with a target audience of foreign nationals abroad is also expected to manage a public relations campaign for the State Department, aimed at domestic audience. Building information tools on the same subject for two different audiences is a difficult balancing act. To make it worse, the Smith-Mundt Act actually made it illegal to conduct perception management activities on the American people and directed that separate budgets exist for the conduct of perception management and public affairs. So not only do these staffs have to differentiate products between their different audiences, but they also need to do so under separate operating authorities and budget tasking. This was too much to ask, and so a decision was made to establish a new organization, the United States Information Agency (USIA), whose sole purpose was to serve as a public diplomacy arm that could in fact conduct these activities legally, but only abroad and then only against foreign citizens.[11]

Thus for almost fifty years, the USIA was the main organization responsible for the conduct of public diplomacy by the federal government. Formed in 1953 under Reorganization Plan No. 8 of the Smith-Mundt Act, this new activity encompassed most of the information programs of the State Department at that time. The lines of authority for this new agency were unique, not only because it operated as an independent organization, with the director of USIA reporting to the president through the NSC, but also because he coordinated his own separate budget. These factors, and resentment of their freedom would become major causes for later reorganization efforts by the State Department over the next forty-five years.

The USIA was involved in support the achievement of strategic national objectives for the course of its existence. Although presidential staffs came and went, the mission remained the same: to promote perception management as a program of the United States. There have also been attempts to strengthen their perception management tools. Specifically, on 6 March 1984, the Reagan administration published a policy titled National Security Decision Directive (NSDD) 130, *U.S. International Information Policy*.[12] This document was envisioned to be a strategic instrument for shaping fundamental political and ideological trends around the world on a long-term basis and ultimately affecting the behavior of governments.[13] Recognizing that a strong international interagency capability for the coordination of perception management was needed, NSDD 130 was follow up to NSC 4 and a predecessor to the Clinton-era Presidential Decision Directive (PDD) 68, *United States International Public Information*.[14]

Throughout its existence, political tensions regarding the proper role and requirement for the USIA led to occasional calls for its disbandment as a discrete agency and integration with the State Department. From 1992 onward, with the end of the cold war, USIA budgetary allocations were reduced, on the basis that the principal function of the USIA—countering Soviet propaganda—no longer existed. From the mid-1990s, the calls for the closing of the USIA became louder, ultimately leading to its disbandment in October 1999, and the transfer of the perception management function back to the State Department.[15]

Proponents of the integration had argued that moving the perception management function to the State Department would allow for a closer and more effective integration between perception management activities and the overall conduct of U. S. foreign policy. Critics feared that this relocation would lead to the dilution of the role in favor of "real" diplomacy. But occurring, as it did, in the last lame duck months of the Clinton administration, any serious decisions regarding the future directions and priorities for U. S. perception management activities were effectively put on hold until a new administration was in place.[16]

Historical Background of Perception Management Arrangements in Australia

The origins of Australia's formal perception management policies also date back to the World War II. A few days after the outbreak of World War II in 1939, the Department of Information (DOI) was established by cabinet directive.[17] The principal function of the DOI was to provide factual and up-to-date information about Australia for dissemination overseas, through both publications and radio broadcasts.[18] In practice, these activities were directed at countering Axis propaganda in enemy, occupied, and neutral states, largely in the Pacific region. These information activities were complemented by military information activities undertaken by the Far Eastern Liaison Office (FELO) and other specialist intelligence and special operations units. At the conclusion of hostilities, FELO (and associated organizations) was disbanded but the DOI continued its activities into the early postwar era, albeit at a reduced level of activity.[19]

Following a change of government, the DOI was dissolved in 1950, to be replaced by the Australian National Information Bureau (ANIB) within the Department of the Interior.[20] ANIB inherited the functions of the previous DOI and its structure, less the shortwave radio broadcasting element (now known as Radio Australia, which became part of the Australian Broadcasting Corporation, ABC).[21] This was the first of a number of occasions upon which Australia's principal perception management organization was renamed and/or administratively moved between government departments.[22] This also represented the beginning of a general trend over the next four decades in which the bureaucratic prominence of, and resources devoted to, Australia's perception management activities gradually declined.

In 1987, the Overseas Information Branch (OIB), as it came to be known, became part of the newly integrated Department of Foreign Affairs and Trade (DFAT), as one element of a broader restructuring of federal executive government.[23] This move represents the first merging of Australia's perception management and foreign policy functions, and offered [24] the prospect of enhancing Australia's perception management activities within this new, broader context of foreign policy. But quite the reverse came to pass, and Australia was unsuccessful in projecting a favorable image of itself overseas. The progressive sidelining of the perception management function within DFAT over the following nine years resulted from a number of interrelated factors. In the minds of OIB personnel, the issue was the prevailing attitude among senior DFAT staff that perception management was peripheral to "real" diplomacy undertaken by "real" diplomats.[25] This attitude tended to marginalize the OIB in bureaucratic terms and to restrict promotion and posting prospects for OIB staff. Further, resources that had been previously devoted to OIB's perception management mission increasingly were diverted toward either domestic public affairs tasks or cultural diplomacy activities that were more palatable to DFAT.[26] This steady, general decline of Australia's perception management capability continued until early 1996.

In April 1996, one of the first decisions made by John Howard, the newly elected prime minister, was to disband the OIB during an overall government cost-cutting exercise.[27] While these cuts affected other sectors of DFAT (as well as other government departments), the fact DFAT decided which programs would be cut fueled the suspicion that agendas dating back to the integration of OIB into DFAT in 1987 were finally being played out.[28] By the end of June 1997, OIB had been dissolved, and its role of coordinating Australia's perception management effort was eliminated, on the basis that this function could be undertaken through the geographically oriented diplomatic divisions of DFAT.

The government also proposed, based on the same cost-cutting plans, that ABC should cease its foreign broadcast services (Radio Australia and now including an international television service, Australia TV) and concentrate solely on domestic broadcasting.[29] Ultimately, Radio Australia continued operations (with significantly reduced resources and broadcasting range) and Australia TV was privatized. But overall, there was concern expressed in the media and by academia that

Australians were concerned that the crippling of perception management capabilities, on the pretext of government fiscal responsibility, would damage Australia's international reputation and standing.[30] In short, it appeared that Australia had lost an image war in the region.

These concerns soon appeared to be founded. In the late 1990s, Australia's image in the region dropped to arguably its lowest ebb since the end of World War II.[31] The progressive decline of Australia's image within the region stemmed from three key factors: long standing views in Southeast Asian states that Australia was in—but not of—the region, the perceived anti-Asian rhetoric of the newly elected One Nation Party politician Pauline Hanson (which was seized upon as evidence of Australia's outsider status), and a number of regional crises—the most significant of these being the Asian financial crisis and the political turmoil in Indonesia (which resulted in the Australia-led military mission in East Timor)—for which Australia (and indeed the West in general) represented a convenient scapegoat.

In response to Australia's increasingly poor image in Asia, DFAT established the Images of Australia unit to counter negative publicity in the region in August 1997.[32] But the miniscule resources provided to the unit—one full-time staff member—were markedly inadequate for the task of repairing Australia's image in the region, and gave little indication that the Australian government was taking perception management matters seriously.[33] So by 1999—the very point at which Australia's image in the region was most seriously tarnished by adverse media reporting—the Australian government had largely undermined its ability to coordinate and disseminate positive messages regarding Australia and its policies, having eliminated the means to do so.

The decline in Australia's ability to project a favorable image in the Southeast Asian region was sharply demonstrated during Australia's leadership of the International Force in East Timor (INTERFET) between late 1999 to early 2000. To give one notable example, media footage of Australian troops disarming anti-independence militia forces that was favorably received domestically (and in Europe and the United States), was used in Southeast Asia to justify official accusations that Australia had acted in an aggressive and racist manner in intervening in East Timor.[34] This occurred in spite of a United Nations mandate for the INTERFET coalition, and an official invitation from Indonesia (then the ruling power in East Timor) to do so. Worse still, officials in Southeast Asian states that had contributed troops to the INTERFET force made such claims.[35] But the absence of effective mechanisms for conducting perception management meant that such charges went unanswered. In short, it appeared Australia had an image war in the region, though this was disputed by DFAT.[36]

Perception Management Practices in the United States since 2001

Initial efforts to reform United States perception management practices commenced in early 2001 after the assumption of power by the new Bush administration. While the Clinton-era PDD 68 was rescinded by the new administration, elements of the perception management coordination arrangements enacted by PDD 68, located within the State Department, remained in place.[37] A report developed by perception management professionals released by the Defense Science Board (published after the 11 September 2001, attacks, but essentially completed prior to the attacks) recommended a range of measures to enhance U. S. perception management policies and activities.[38] The report made specific recommendations regarding the requirement for a National Security Presidential Directive (NSPD, the new Bush administration's successor to the Clinton administration's PDDs) on perception management and proposed specific coordination measures.[39] While none of the recommendations of the report were ever specifically implemented, the report did, to an extent, inform the changes made to United States perception management practices after 11 September 2001.

In the immediate wake of the 11 September 2001, attacks, the U. S. perception management effort appeared somewhat rudderless. The Office of International Information Programs (IIP), as the remaining elements of USIA were now known, within the State Department looked to the National Security Council (NSC) to provide direction for perception management efforts in the new global security environment. But given the rescinding of PDD 68 in early 2001, and the absence of a new presidential directive regarding perception management, such direction was not forthcoming. Unsurprisingly, this period was also occasioned by several perception management missteps: the war against al Qaeda and the Taliban was characterized as a "crusade," and the military campaign against them was initially dubbed "Infinite Justice." While such labels resonated with the domestic audience, they were profoundly alienating to Islamic audiences across the globe.

In response to the apparent perception management policy vacuum, the Department of Defense (DoD) initially contracted the Rendon Group—a civilian company that specializes in strategic communications advice—to assist with a perception management campaign centered on military operations in Afghanistan.[40] Following its work with the Rendon Group, DoD established the Office of Strategic Influence (OSI) for the conduct of more general perception management activities in support of the war on terrorism.[41] But from the outset, the OSI was undermined by concerns—in part generated by DoD and State Department public affairs officials—that it would disseminate disinformation to foreign and domestic media outlets. In such an environment it was impossible for the OSI to operate effectively, and it was disbanded in February 2002.[42]

In spite of the above missteps, some improvement in U. S. perception management practices was evident by late 2001 and early 2002. The White House established (with coalition military partners) a network of Coalition Information Centers (CIC) linking Washington, London, and Islamabad, which were able to work in synch with the news cycles of key Middle East media outlets. Accordingly, they were well situated to respond to a dynamic operational environment, and to counter Taliban and al Qaeda disinformation and propaganda.[43] Despite their success, the CICs were regarded as a temporary response to the particular requirement of concerted military operations against the Taliban and al Qaeda in Afghanistan. As these came to an end in mid-2002, the CICs consequently were shut down.

Just as the CICs were ceasing operations, permanent arrangements to undertake the role performed by the CICs were put into place. In June 2002 (though not formally established by Executive Order until January 2003) the White House established the Office of Global Communications (OGC) to coordinate global perception management activities.[44] And in September 2002, the NSC established a Strategic Communication Policy Coordinating Committee (PCC) to strengthen interagency mechanisms for the development and dissemination of messages to international audiences.[45] But while these new arrangements were announced with great fanfare, and at least appeared to have the potential to enhance U. S. perception management activities, they ultimately failed to live up to their promise.

Although the OGC was established to coordinate the conduct of a strategic long-term perception management campaign, the activities undertaken could be characterized as tactical and reactive. The Strategic Communications PCC only met on a handful of occasions, achieved little, and was regarded as moribund by 2004.[46] But most important of all, the fundamental credibility of U. S. perception management efforts was undermined in the wake of the war in Iraq in mid-2003, by both the conspicuous absence of Iraqi weapons of mass destruction, and the highly publicized mistreatment of prisoners in Coalition custody.

In response to growing concern about the effectiveness of U.S. perception management efforts, the White House acted to address their perceived failings. In September 2005, Karen Hughes (who had previously been involved with the OGC in 2002–2003) was appointed as the Undersecretary of State for Public Diplomacy and Public Affairs in the State Department.[47] While it was generally recognized that the appointment of Hughes represented strong commitment by the Bush administration to enhance perception management activities, Hughes' background in domestic public relations rather than perception management raised concerns about her suitability for this appointment.

At around the same time, the OGC was disbanded and incorporated into the newly created Office for Strategic Communications and Global Outreach within a reorganized NSC.[48] At time of writing, it is not clear what effect these new arrangements will have on the conduct of perception management activities. But what is certain is that the United States currently faces an image problem of literally global proportions. The 2006 Pew Global Attitudes Report indicated that overseas public support for the United States, both generally and in the context of the struggle against al Qaeda, had declined significantly, even amongst the citizens of allies.[49] So while there had been some reforms to its perception management arrangements, in mid-2006 the United States was faced with its greatest perception management challenge since the 11 September 2001 attacks.

Perception Management Practices in Australia since 2001

Efforts to improve the effectiveness of Australia's perception management were undertaken in mid-2000. To a considerable extent, the rationale for these measures arose out of lessons drawn from Australia's experience as the INTERFET coalition lead nation. In terms of the organizational arrangements for conducting perception management, the Images of Australia unit in DFAT was expanded, ultimately to become the Images of Australia Branch (IAB) in 2001. In terms of the means for projecting favorable images of Australia, the Minister for Foreign Affairs and Trade announced in June 2001 that the ABC would commence a new international television service—ABC Asia-Pacific—in early 2002.[50] This effectively replaced the service that had been provided by the privately owned Australia TV, which had ceased operations in March 2001.[51]

While these initiatives improved Australia's ability to conduct perception management, it can be argued that such an improvement largely represented an undoing of the neglect that Australia's perception management capability had suffered over the preceding decade. By expanding the Images of Australia unit to branch size, the status of IAB (in broad organizational terms) within DFAT was now on par with that previously held by OIB prior to its disbandment back in 1996. And the new international television service bore a close resemblance to the previous Australia TV international service—or at least to the appearance of Australia TV prior to its privatization in 1997. Further, the questionable quality of much of Australia TV's programming after privatization had arguably undermined its effectiveness as a means for projecting a favorable image of Australia to the region during its period of operation. So these initial measures for enhancing Australia's perception management capability involved little that was genuinely new.

It was not until after the Bali bombings of October 2002 that the potentially lethal consequences of being regarded as a scapegoat—and therefore the significance of perception management to the conduct of national security policy—were thrown into sharp relief for Australian policy-makers.[52] Prior to the bombing, a number of al Qaeda statements had specifically identified Australia as a "crusader state"—and therefore as an appropriate target for terrorist acts in the name of furthering its vision of Islam.[53] Specific mention was made of the role of the Australian military in the East Timorese independence process: "they landed to separate East Timor, which is part of the Islamic world."[54] Following the bombing, another al Qaeda statement (allegedly made personally by bin Laden) drew a direct connection between Australia's role in East Timor (and later in Afghanistan) and the Bali bombings.

It is important to note that the terrorist group that conducted the Bali bombings—Jemaah Islamiyah—is inspired by and affiliated with, rather than directly controlled by, al Qaeda. So the propaganda statements of al Qaeda noted above should not necessarily be seen as a direct call to action for Jemaah Islamiyah against Australian targets. But propaganda of this nature (be it from al Qaeda or from Jemaah Islamiyah itself) has the potential to goad some individuals already antipathetic toward Australia to conduct terrorist acts. And while the Bali bombings did not seek to target Australians exclusively, it was inevitable and known that Australians would be the principal victims of the attacks.[55] From these premises, it followed that efforts to project a positive international image of Australia should constitute an integral part of Australia's overall approach to national security, including it strategy to undermine the wellsprings of transnational terrorism.[56]

In the years following the Bali attack, the prominence of perception management as an instrument of foreign policy and the resources devoted to it by government increased. The 2003 DFAT White Paper, *Advancing the National Interest*, contained extensive discussion of the significance of projecting a positive image of Australia to the world, and the various means for doing so.[57] This contrasted markedly with the 1997 DFAT White Paper, *In the National Interest*, which touched only briefly on perception management and in doing so focused almost exclusively on cultural diplomacy.[58] The 2004 DFAT White Paper specifically identified the "battle of ideas" as a key element of the overall strategy to counter transnational terrorists.[59] So by the midpoint of the first decade of the twenty-first century, the prominence of perception management as an instrument of Australian national security policy was arguably at its highest level since World War II.

But even though the prominence of perception management within Australian national security policy had increased, the organizational framework at the national level within which it took place remained the same both before and after 2002. No new arrangements for the specific coordination of perception management activities across government were established. Rather, perception management issues continued to be considered via DFAT membership of a range of specially convened (rather than standing) inter-departmental committees and task forces (at the working level), and the standing Secretaries Committee on National Security and National Security Committee of Cabinet (at the senior officer and ministerial levels). This reflected the organizational structure that had been in place since the election of the first Howard government in 1996, which in turn was essentially a slightly modified, post–World War II national security structure.[60]

At time of writing, there are no standing high-level policies in Australia that coordinate perception management activities across the government. In this respect, perception management differs somewhat to other elements of national security (specifically, counterterrorism activities and critical infrastructure protection), where there has been at least some measure of organizational change since 2001 to coordinate action between different government departments and agencies.[61]

But while the profile of Australia's perception management efforts has increased, there is little real understanding of how effective (or ineffective) these activities have been. There have been no mechanisms put in place to evaluate systematically the effectiveness of Australia's perception management policies and efforts.[62] And in the absence of such evaluation mechanisms, it is difficult to focus or enhance Australia's perception management activities. So while there is some sense that Australia's perception management practices are more effective now than they were prior to 2001, there is little unambiguous evidence to support this belief.

Comparing the Perception Management Experiences of the United States and Australia

A comparison of the experiences of the United States and Australia with respect to their perception management practices reveals both similarities and differences prior to 2001 up to the present. The most obvious similarity between the two states with respect to perception management is the decline in the prominence of and resources devoted to perception management in both states up to 2001, followed by something of a renaissance thereafter. In both states, perception management activities were scaled back prior to 2001, both because of the belief that their relevance had declined in the post–cold war period, and because of domestic political agendas. In the wake of the 11 September attacks, both states experienced renewed recognition of the value of perception management as an instrument of national power. This was ultimately reflected in an increased profile for, and increased resource allocation to, perception management activities in both states after 2001.

Two other similarities regarding perception management in the United States and Australia also stand out. First, in both states the integration of the perception management function into the broader foreign policy bureaucracies had generally negative consequences for the conduct of perception management activities. This was most notable in Australia, where the integration of Australia's perception management practices into the foreign policy bureaucracy led to its effective elimination, prior to being reconstituted in 2001. And while the disbandment of USIA in 1999 did not eliminate formal procedures for the conduct of perception management in the United States, it arguably did have a negative impact on their effectiveness.

Secondly, in both states the effective conduct of perception management activities was at times undermined by the dissemination of themes and messages that, while attractive to domestic and aligned audiences, were profoundly alienating to broader international audiences. In the case of both states, one apparent reason for such perception management missteps was indifference to or ignorance of the differences between domestic and international target audiences. Or to put this matter in other terms, the difference between the conduct of public affairs and perception management activities. This is not to say that public affairs and perception management activities should (or in the Information Age even can) be conducted in isolation of one another. But international audiences are not the same as domestic ones, and treating them as if they are can have serious negative consequences.

The most obvious difference between the perception management practices of the United States and Australia (both before and after 2001) is their scale and scope. But this difference is a scarcely remarkable consequence of the different interests of a superpower as compared to a medium-power state, and warrants no further consideration. However, significant differences between the United States and Australia do exist in the organization of perception management. While there was considerable organizational change in perception management arrangements in the United States after 2001, there was little organizational change in Australia over the same time period. This lack of change, also seen in counterterrorism and critical infrastructure protection, reflects a broader cultural difference between the two states. There is a higher level of procedural and organizational informality in the Australian political system.[63]

Concluding Remarks: The Significance of Perception Management in Fighting Transnational Terrorist Organizations

This chapter opened with a brief discussion of the implications of the Information Age for the conduct of perception management activities. In particular, it was noted that reach, access, and availability of contemporary information and communications technology makes perception management a particularly effective and attractive instrument of power for transnational terrorist groups. The extensive employment of a variety of information and communications technologies to disseminate propaganda globally by al Qaeda and its affiliates is readily evident, and will not be further discussed here. Rather, these concluding remarks will address the role of perception management as an instrument of power employed by states in the struggle against transnational terrorists.

Given the extensive use of propaganda by al Qaeda and its ilk, it thereby follows that that one of the key roles for perception management in the fight against transnational terrorist groups is countering the propaganda and misinformation disseminated by such groups. But in addition to countering terrorist propaganda, there is subtler, yet arguably more important, role for perception management activities in the battle against global terrorist organizations. This lower-key role involves the general building of understanding and good will toward the United States (and its allies) within the general populations of current and potential partner states in this global campaign. Success in the struggle against transnational terrorism is critically dependent upon international cooperation—an issue recognized in the counterterrorism strategies of the United States and Australia.[64] Such cooperation is enacted via formal bilateral and multilateral agreements between states pertaining to a wide range of issues such as intelligence, military and law enforcement operations, aid and development assistance, and other policy areas.

These formal government-to-government agreements (and the cooperative activities that take place as a consequence of such agreements) are ultimately founded on a basic level of acceptance, if not actual support, in the general populations of partner states. So arguably the most important role for perception management activities in the global struggle against transnational terrorism is the establishment and maintenance of good will and understanding toward the Western world in the general populations of current and potential partner states. For it is only when there is at least some measure of popular good will and understanding that there will be acceptance and support for cooperation with the United States and its allies against transnational terrorist organizations. And in the absence of popular acceptance and support for such measures in partner states, international cooperation against terrorist groups will ultimately evaporate.

The building of understanding and good will in the general populations of partner states through perception management activities is critically dependent upon several key requirements.It requires access to the intended target audiences through appropriate means for information dissemination. It requires an understanding of the beliefs and values of the target audiences, and how these beliefs and values effect the audiences' interpretation of perception management messages.[65] But most of all, it requires that the messages received by target audiences are credible and are consistent with the reality that they experience.[66] This argues for close linkages between perception management activities and the overall conduct of foreign and national security policies.

In part, this requisite link reflects the imperatives of consistency: messages that do not accord with the experienced reality of target audiences are unlikely to be accepted. How global audiences perceive the objectives and conduct of foreign and national security policy must also be considered. In this regard, Joseph Nye has noted that "the best advertising can't sell an unpopular product."[67] When the objectives and conduct of foreign and national security policy are of themselves a cause of negative foreign public opinion, it is unlikely that perception management activities alone can be successful in changing these opinions.[68] Any approach to perception management, or indeed the conduct of foreign and national security policy as a whole, that ignores this reality courts failure of the worst sort.

CHAPTER 10

Conclusion

Leigh Armistead, Edith Cowan University

We now live in an information age, which has been called an era of net
works. Loudly proclaimed by many throughout the world, it is interest
ing to compare and contrast the differences between informational rhetoric and reality, especially in the employment of IO, as has been attempted in this
book. A relatively newly defined activity, this transformation of traditional uses of
power promises to revolutionize the manner in which warfare, diplomacy, business, and a number of other activities are conducted. However the gap between
proposed capability and actual conduct of operations in the United States government is wide and while strategic doctrine and guidance may exist to best utilize
the power of information, in fact, actual information campaigns are almost always
conducted at a tactical level. In this book, the authors develop definitions and
models that articulate not only why this divide exists, but also specific strategies
for utilizing IO in a manner that best optimizes the inherent capabilities of this
element of power. It is hoped that the conclusions developed in this book may be
useful for future IO planners, as well as senior-level decision makers.

Information is and always has been a somewhat a nebulous term, but in this
new era it possesses a capability that is now considered crucial to the success of
U.S. national security. However many questions arise regarding the best method
for successfully utilizing the power of information to the fullest extent by the
United States government. Because IO crosses so many boundaries within the
interagency processes, it is often very difficult to quantify exactly what constitutes
an information campaign. This is because you now have other organizations within
the federal bureaucracy, such as the State Department, which have traditionally
concentrated on diplomatic efforts but are now being asked to facilitate strategic
IO activities around the world. Not only is this kind of tasking abnormal for these
different cabinet agencies, but it also belies their normal chains of communication
and day-to-day procedures. So more often than not, the most recent attempts to
conduct strategic high-level IO activities in the United States are aborted for a

more tactical set of options that are normally conducted by the Department of Defense as part of their standard set of operations.

A good example of these dichotomies is seen in the three military activities conducted most recently by U.S. forces in the last decade. Whether it was Kosovo, Afghanistan, or Iraq, the primary focus of these campaigns from the viewpoint of Washington, D.C. was on military victory. In none of these operations did IO play the transformational role that its advocates have predicted, and while a number of capabilities and related activities have been utilized, often with good success, at best these efforts are still almost entirely concentrated at the localized or tactical level. Nowhere has the strategic revolution in warfare advocated by informational power enthusiasts materialized.

Yet these transformational ideas are crucial, because as the events of 11 September 2001 indicate, military, political, or economic power are often simply ineffective in dealing with these new kinds of threats to the national security of the United States. The aforementioned terrorist attacks were a blow to the American public and government images that affected the perceptions of many Americans. In addition, as the text indicated, the fear produced by the terrorist acts can only be defeated by using a comprehensive plan in which information is a key element. For as both John Arquilla and David Ronfeldt argue in the future networks will be fighting networks. Both Operation Enduring Freedom (OEF) and Operation Iraqi Freedom (OIF) were such examples, networks in the form of information campaigns fought against networks made up of perceptions, and the side that will ultimately emerge as the victor is the one that can best shape and influence the minds of not only their adversary, but their allies as well.

So why in this post–cold war era, when the greatest threats known to mankind such as nuclear war are lessened, is the United States still under attack from a number of different enemies, including the al Qaeda terrorist network? Once again, there are many reasons, but primarily it is because the perception of the enemy has changed. While there are still "rogue states" (in U.S. terms) that can occupy the politicians and give credence to budget appropriations, other groups including extremist religious factions are freer to operate and to carry out attacks on the United States, in this post-bipolar period. Most of these nongovernment organizations (NGOs) or terrorist groups are no longer operating underneath the umbrella of a superpower and therefore have much more autonomy than ever before. Over the past fifteen years, and especially in the last few, there has been an explosion of attacks on the United States, in some of which information has played a key role. While individuals conducted a number of these incidents, others were the work of activists, foreign military units, terrorists, and even nation-states. Thus, we are now at a point where each of these terrorist operations highlights the vulnerability of America and its population to these new types of warfare, where information and the integration of the government play a key role. Yet as mentioned previously, there is tremendous gap between the theoretical potential of IO and its day-to-day implementation.

Likewise, the overall the hope is that changes to the DoD intelligence structure will bolster the ability of agencies to support the IO capabilities of the United States government. As has been enumerated from the earliest policy development of Command and Control Warfare (C2W) in the post–Desert Storm era, intelligence was always designed to be the bedrock on which IO resides. Without good intelligence, IO is nothing, and so it is imperative that any reorganization or change in structure must lead toward a more informed and capable agency. In conclusion then, while there has been a number of changes to IO policy and organizations within the U.S. government over the last couple of years, most seem to be more of an operational effort to make IO more useful. The radical and revolutionary aspects of this warfare area appear to have abated, as we separate the hype from the reality of IO, and the Wild West mentality has been replaced by a more operational viewpoint. So in the end, it may in fact be that the normalization or suburbanization of IO over the last few years that may be the most important point of this book.

The interagency organizations of the U.S. federal government have also recognized the difference between IO capabilities and actual practices, and are attempting to better utilize information as an element of power. Evidence of this can be seen in the creation of the Department of Homeland Security, reorganization changes at the State Department, and the attempts to train an IO educated workforce at authorized academic centers of excellence. In addition, by managing information and using planning tools to synchronize, synergize, and neutralize influence-based activities in an overall plan to effect the adversary, officials in Washington, D.C. have also attempted to enable the horizontal integration of these activities across the whole interagency and coalition environment. A good example of this was the effort to coordinate perception management messages of the United States, NATO, and the United Kingdom during the Kosovo conflict in 1999.

This shift from one era to another is not without precedent. The United States became a world leader in the industrial age because it could mobilize the collective might of its population through mass production, automation, economic incentives, and geographic location. However, this era of industrial might is in decline and the information revolution is now upon us.[1] The ability for the United States government to conduct influence campaigns around the world is under a tremendous amount of stress and uncertainty. In previous generations, the practitioners could count on a monolithic enemy such as the Soviet Union and somewhat static communications technology throughout the cold war (broadcast network TV and radio). This situation unfortunately created an erroneous belief that one could control the information that was broadcast to the known adversary, but this is no longer possible in today's environment. The Internet and other emerging communication networks (wireless, peer-to-peer, etc.) have forever destroyed the power formerly resident only in the government, and that asymmetry now gives the power of information to all. This is a good example of the power of bottom-up execution and control. Alvin Toffler and Heidi Toffler alluded to this capability in their book *War and Anti-War*, when they discuss the decline of the mass media and the ability to compartmentalize influence campaigns. In addition, while federal bureaucrats could at one time count on the fact that they owned or could somewhat control the dissemination outlets for information, this is increasingly no longer the case. The use of websites, blogs, streaming video, portals, and other alternative news sources has ended the government's monopoly of information control.

This concept of the power and control of information is at the core of this book. While most of the analysis focuses on the key area of perception management (PM) within IO, it is often referred to by many different monikers depending on which branch of the USG is being referenced. Some of these names are psychological operations (used by the DoD), public diplomacy (DoS), strategic communications (NSC), and influence operations (White House). Yet in essence all of these terms can be considered analogous and in this text, the authors have elected to use these terms somewhat interchangeably. In addition, while there are many other capabilities of IO, such as deception, electronic warfare, and the like that could also be examined, the authors instead chose to narrow the subject to this portion of IO. This is because it is during attempts to conduct these kinds of influence campaigns that the United States government has had the most difficulties recently, and where the divide between theory and reality is the greatest. In the authors' opinions perception management offers the most potential for change within the federal bureaucracy.

This book is therefore an attempt to bring together the disparate efforts by various federal organizations, and define IO capabilities, with an emphasis on perception management. In addition, this research has attempted to investigate how the key agencies of the U.S. government can use the inherent power of information to better conduct strategic communication and influence campaigns in the future. This book also attempts to use these campaign models to create a strategy for U.S. utilization of IO. The authors hope that this book will illuminate a process that can be used to transform federal organizations and allow them to better understand and use the power of information to meet the future threats.

The bottom line is the question of whether the federal government can conduct an effective information campaign in this changing environment, while assuring the security of its networks and information systems in this new architecture. To do this, will the United States need to change its collective interagency structure that has evolved over two hundred years into a more networked organization that can master the issues in the information age? And finally, will the United States remain a dominant player during this new era, when industrial capacity is not nearly as important to a nation as its interconnectivity of information nodes? This book has endeavored to answer these questions while producing a better strategy for the United States to best conduct information operations in this new era.

Notes

Chapter 2: Information Operations: The Policy and Organizational Evolution

1. Prepared Statement For the House Budget Committee on the FY 2004 Defense Budget Request Remarks as Prepared for Delivery by Deputy Secretary of Defense Paul Wolfowitz, Washington, DC, Thursday, 27 February 2003.

2. This terminology comes from the SecDef's "Information Operations Roadmap", published on 30 October 2003. While the entire document remains classified, the greater part of it was released to the public in 2006 in a redacted format and is available on the Internet. The authors of this chapter feel that it was a strange and inexplicable mistake to omit outer space from the list of operational realms, because warfare and other military operations are now conducted seamlessly across five physical domains: air, land, sea, outer space and cyberspace.

3. The NMS-CO was still classified at the time of this writing and unavailable to the general public, although major segments of it are unclassified. This author was part of the team that produced it, my major contribution being the definition of cyberspace.

4. U.S. Department of Defense Instruction 5200.40. DoD Information Technology Security Certification and Accreditation Process (DITSCAP). (31 December 1997).

5. Interview with Dr. Corey Schou, 21 February 2004.

6. Corey Schou, W. V. Maconachy, and James Frost," Organizational Information Security: Awareness, Training and Education to Maintain System Integrity," In Proceedings Ninth International Computer Security Symposium, Toronto, Canada, May 1993.

7. U.S. Commission on National Security/21st Century, Road Map for National Security:Imperative for Change. (15 February 2001).

8. Although the original IO Roadmap was classified, a heavily redacted version was released in 2005 and is available on the Internet. Our summary in chapter 2 is from pages 6-7 of the Roadmap. Also see Chris Lamb's article on the IO Roadmap in Joint Force Quarterly (Issue 36, December 2004), 88-96; available online at www.ndu.edu/inss.

9. See pp. 10-15 of the Roadmap, from which the recommendations discussed above were drawn.

10. Roadmap, p.11

11. Joint Publication 3-13, Information Operations, 13 February 2006. The definition of information superiority used in here—"the operational advantage derived from the

ability to collect, process and disseminate an uninterrupted flow of information while exploiting or denying an adversary's ability to do the same"—is unfortunate and is vastly different from the way an airman would define air superiority or the way a seaman would define naval superiority, which always comes back to the ability to use the environment while denying an adversary the same degree of utility. Although by 1945 the U.S. had air superiority over Europe and naval superiority in the Atlantic, we lost airplanes and ships until the end of the war. Superiority is a measurement, not an absolute.

12. For a brief discussion and analysis of each see the U.S. Army War College publication "Information Operations Primer," published in January 2006.

13. John Bennett and Carlo Munoz, "USAF Sets up First Cyberspace Command," in Inside Defense.Com NewsStand (4 November 2006).

14. Memo from Karen Hughes, Chair of the PCC on PD/SC to PCC Participants, dated 18 October 2006, subject "National Strategy Plan to Govern US International Public Diplomacy and Strategic Communication."

15. William P. Hamblet and Jerry G. Kline, "PDD 56 and Complex Contingency Operations." Joint Forces Quarterly (Issue 24, Spring 2000).

16. Interview with Jeffrey Jones, 13 August 2003.

17. Interview with Bill Parker, 24 March 2004.

18. Interview with Peter Kovach, 31 March 2004.

19. "Changing Minds, Winning Peace: A New Strategic Direction for U.S. Public Diplomacy in the Arab and Muslim World." Report of the Advisory Group on Public Diplomacy for the Arab and Muslim World (1 October 2003). p.1.

20. Ibid, p. 10.

21. Including this author, as Dan Kuehl was a member of the DSB Task Force and contributed to its report.

22. Therefore in a sense, the DSB serves as a verification of sorts that the research conducted as part of this book is on track with regard to the needs and deficiencies of IO with regard to hype versus reality.

23. John Arquilla and David Ronfeldt wrote several brilliant analyses published by RAND in the 1990s that emphasized this networked element of the strategic environment, including Cyberwar is Coming! and The Advent of Netwar (1996), in *Athena's Camp: Preparing for Conflict in the Information Age* (1997), *The Emergence of Noopolitk: Toward and American Information Strategy* (1999), and Networks and *Netwars: the Future of Terror, Crime and Militancy* (2001).

24. Dr. Tuija Kuusisto, Rauno Kuusisto and Leigh Armistead. System Approach to Information Operations. A paper presented at the European Conference on Information Warfare (July 2004).

25. Quadrennial Defense Review Execution Roadmap for Strategic Communication, 25 September 2006.

26. U.S. National Strategy for Public Diplomacy and Strategic Communication (draft), circulated under cover of a memo from Karen Hughes, Chair for the PCC on Public Diplomacy and Strategic Communication to PCC participants, 18 October 2006.

27. John T. Benett, "USAF Sets up First Cyberspace Command," InsideDefense.Com, 4 November 2006.

29. Interview with Jeff Jones, 17 April 2002.

30. Interview with John Rendon, 18 April 2003.

31. Alter, Jonathan, "Why Hughes Shouldn't Go." *Newsweek*, 6 May 2002, p. 49.

32. See the work of James Guirard, who has been arguing this issue for nearly a decade. Also see recent work of Doug Streusand, on the faculty at Marine Corps University

at Quantico, and Army Colonel Harry Tunnell (one of my former students at the National Defense University), who have raised this issue within Defense circles.

33. Franklin Foer, "Flacks Americana—John Rendons Shallow PR War on Terrorism." *New Republic*, 20 May 2002.

34. Interview with Chris Pilecki, 15 April 2002.

35. Florian Rotzer, "Schon Wieder Eine Chinesische Mai-Offensive?" *Die Heise*, 25 April 2002, accessed at www.heise.de/tp/deutsch/special/info/12401/1.html.

36. Interview with Tom Timmes, 15 April 2002.

37. Thomas Ricks, "Downing Resigns as Bush Aide." *Washington Post*, 28 June 2002, accessed at www.washingtonpost.com/wp-dyn/articles/A56316-2002Jun27.html.

38. Bob Woodward and Dan Balz, "Aus Trauer Wird Wut," *Der Welt*, 16 February 2002, accessed at www.welt.de/daten/2002/02/16/0216au314625.htx?; Rötzer, Florian, "Aus für die Propaganda-Abteilung des Pentagon," *Die Heise*, 21 February 2002, accessed at www.heise.de/tp/deutsch/special/info/11952/1.html; Florian, "Rumsfield: Pentagon Lügt Nicht," *Die Heise*, 21 February 2002, accessed at www.heise.de/tp/deutsch/special/info/11895/1.html; Creveld, Martin van, "Die USA im Psychokrieg," *Der Welt*, 03 March 2002, accessed at www.welt.de/daten/2002/03/01/0301fo317407.htx?

39. U.S. Department of Defense, Office of the Under Secretary of Defense for Acquisition, *Technology & Logistics*, Report of the Defense Science Board Task Force on Managed Information Dissemination, October 2001.

40. Bruce Gregory. Interview with author. Adjunct Professor, George Washington University, 16 April 2002.

41. Hoffman, David, "Beyond Public Diplomacy." Foreign Affairs, March/April 2002.

42. Before one thinks that all of these efforts are sinister propaganda campaigns, one example of an OSI mission that had been proposed included global education, an issue that is very prominent in the draft new National Strategy for PD/SC. OSI developed a plan to influence some of the madrassas, the Islamic schools that exist across much of the Arabic world and are places that are widely seen as incubators for radical Islamist indoctrination.

42. To do this they wanted to provide more information about the outside world, including Internet access, adopt a school, and establish foreign exchange programs in order to broaden the students' horizons. This idea of globalism, that is, it is the role of DoD to educate a Pakistani child, might seem a bit of a stretch, and maybe that is where Clarke objected, but on a deeper level, these are exactly the types of operations that need to be undertaken by some organization within the U.S. government

43. Ward, Brad, LTC (USA). Interview with Author. Department of Defense Representative to the IPI Core Group at the Department of State.15 June 2001 and 15 April 2002.

44. Interview with Brad Ward, 15 June 2001.

45. Jeff Jones. Interview with author, 17 April 2002, Washington, DC.

Chapter 3: Perception Management: IO's Stepchild

1. Alan D. Campen, ed., "Communications Support to Intelligence," in *The First Information War: The Story of Communications, Computers, and Intelligence Systems in the Persian Gulf War* (Fairfax, VA: AFCEA International Press, 1992), 53.

2. Department of Defense, Joint Publication 3-13, *Joint Doctrine for Information Operations*, Washington, DC: 9 October 1998, accessed at http://www.dtic.mil/doctrine/jel/new_pubs/jp3_13.pdf.

3. Department of Defense, Joint Publication 1-02, *Dictionary of Military and Associated*

Terms (as amended through 17 December 2003), accessed at http://www.dtic.mil/doctrine/jel/doddict/index.html.

4. In this vein, for example, one could look to the U.S. Institute for Peace's Cross-Cultural Negotiation Project for examinations of other nations' diplomatic approaches as a tool toward understanding how "others" perceive differently than what might be expected from an "American" point of view. See, for example, Kevin Avruch, *Culture and Conflict Resolution,* Washington, DC, USIP, 1998; and, Raymond Cohen, *Negotiating Across Cultures: International Communication in an Interdependent World*, Washington, DC, USIP, 1997. For other publications, see www.usip.org.

5. The inclusion of public affairs as a perception management tool is controversial. PA is designed to gain and maintain public support by achieving a high degree of credibility through the timely, accurate, and complete release of information. Many public affairs officers consider that a close association between PA and PSYOP, Intelligence, and deception weakens their credibility and therefore hinders their ability to contribute effectively to mission accomplishment.

6. For more details, see Stephen T. Hosmer, "Psychological Effect of U.S. Air Operations in Four Wars: 1941–1991: Lessons for U.S. Commanders," Santa Monica, CA: Rand Study, Project Air Force, 1996. See also, Stephen T. Hosmer, "Effects of the Coalition Air Campaign Against Iraqi Ground Forces in the Gulf War," Santa Monica, CA: Rand Study, Project Air Force, 2002, 119–137.

7. Perception management thus must be considered a critical pillar of three of the four key DoD objectives as outlined in the *QDR*: to assure allies, to dissuade potential adversaries, and deter adversaries. One could consider successful deterrence (and dissuasion and compellence) as requiring C5:

- capability to act;
- commitment (will) to use that capability if required;
- communication of that capability and commitment;
- comprehension by the potential adversary of the communicated capability and commitment;
- cooperation from the adversary in agreeing that being swayed to act in accordance with one's desires.

Critical to all of this are communication and comprehension—which is PM. Note, one can communicate exaggerated capability and commitment but, like in poker, must be ready to suffer the consequences if one's bluff is called. The author would like to thank Phil Meilinger, of the Northrop Grumman Analysis Center, for the idea of applying the C5 concept in this context.

8. Two days before 9/11, Osama bin Laden arranged to have Commandant Massoud (one of the Northern Alliance's leaders who was quite respected in the West) killed. His calculus might have been that the only way the United States would retaliate to the massive attacks on 9/11 would be by using local surrogates to engage al Qaeda and the Taliban. By eliminating Massoud, he might have hoped to take out the most likely candidate ally of the United States. His calculation proved wrong, as the United States did not respond by anything less than total war against the Taliban and al Qaeda.

9. Readers should refer to current PA and PSYOP doctrines for details.

10. Rick Brennan and R. Evan Ellis, "Information Warfare in Multilateral Peace Operations: A case study of Somalia," report for Science Applications International Corporation, April 1996.

11. Steven Kull and Clay Ramsay, "U.S. Public Attitudes on Involvement in Somalia,"

report for Program on International Policy Attitudes, University of Maryland, College Park, MD, 26 October 1993.

12. Dusan Basic (deputy director of Media Development), interview by the author, Sarajevo, July 1998. Media Development is a Bosnian non-governmental organization monitoring local media and encouraging journalistic professionalism. It is sponsored by the European Commission and the School of Journalism in Lille, France.

13. Rapport spécial sur les médias, rapporteur spécial désigné par la résolution 1994/72 de la commission des droits de l'homme des Nations-Unies [Special report/ratio on the media, special rapporteur appointed by resolution 1994/72 of the commission of the human rights of the United Nations], E/CN4/1995/54, 13 December 1994, 35.

14. Alex Ivanko (former UNPROFOR spokesman), interview by the author, UNMIBH headquarters, Sarajevo, October 1996. Bosnia is certainly not the only place where the international media has been accused of being an instigator of violence simply by its presence. In Haitian and Cuban refugee camps in Guantanamo Bay, the refugees regularly staged demonstrations and became unruly when media appeared. In 2000, the UN asked the international media in Sierra Leone to pledge that they not broadcast "executions" manufactured for their benefit.

15. Material from Bill Arkin of Human Rights Watch, who has conducted an on-the-ground survey of collateral damage claims in Afghanistan. He has numerous examples (with photos and other documentation) of instances where intelligence information could have supported more robust public affairs, public diplomacy, and psychological operations activities. For an interesting discussion of the journalistic problems related to reporting Afghanistan collateral damage cases, see Lucinda Fleeson, "The Civilian Casualty Challenge," *American Journalism Review*, April 2002.

16. Since fall 2001, al Jazeera has gained much fame in the United States through its broadcast of bin Laden tapes. During Operation Iraqi Freedom, al Jazeera's infamy skyrocketed through its broadcast of images of U.S. casualties. For this author's views on the issue, see, "Is U.S. Casualty Reporting Suffering From Double Standards?" *Foreign Policy in Focus*, 15 April 2003, http://www.fpif.org/outside/commentary/2003/0304media.html.

17. See "A New Antisemitic Myth in the Arab Press: The September 11 Attacks Were Perpetrated by the Jews," The Middle East Media Research Institute, Special Report no.6, 8 January 2002, www.memri.org.

18. In the winter of 2003, antiwar protests in London and Rome were some of the largest demonstrations since Vietnam. Note that the United Kingdom and Italy were two of the strongest members of the "coalition" going to war with the United States.

19. According to Aaron Brown, CCN was flocked with letters of protest when it showed pictures of casualties. The network therefore decided to curtail such coverage. Instead, and to retain its credibility with different audiences in the world, CNN international kept covering the issue. However, the discrepancy was not lost on international audiences.

20. After the cold war, Congress decided to cut back funding for the United States Information Agency (USIA) to reintegrate it into the State Department. As a result, programs and broadcasting operations have been cut back.

21. On OSI, see for example: Rachel Coen, "Behind the Pentagon's Propaganda Plan," Fairness and Accuracy in Media, Extra! Update, April 2002, http://www.fair.org/extra/0204/osi.html.

22. In 1950, the Department of State sponsored Project Troy with the charter to determine how best to "get the truth" behind the Iron Curtain. See Allan A. Needell,

"Truth Is Our Weapon: Project TROY, Political Warfare and Government-Academic Relations in the National Security State," *Diplomatic History* 17, no. 3, Summer 1993, 399–420.

Chapter 4: Information Operations in the Global War on Terror: Lessons Learned From Operations in Afghanistan and Iraq

1. General Tommy Franks, "Statement to the Senate Armed Services Committee," 7 February 2002.
2. Department of Defense, *IO Roadmap*, October 2003.
3. http://www.psywarrior.com/terrorism.html, (accessed 13 October 2003).
4. Thom Shanker and Eric Schmitt, "Firing Leaflets and Electrons, U.S. Wages Information War," *New York Times*, 24 February 2004, https://www.us.army.mil/portal/jhtml/earlyBird/Feb2003/e20030224156370.html (accessed 15 July 2003).
5. Senior U.S. Defense Official, "Background Briefing on Enemy Denial and Deception," 24 October 2001, http://www.defenselink.mil/news/Oct2001/g011024-D-6570C.html, (accessed 26 October 2001).
6. Terrence Smith, interviewed by Peter Jennings, "Window on the War," *PBS Online News Hour*, 8 October 2001, http://www.pbs.org/newshour/bb/media/july-dec01/jazeera_10-8.html, (accessed 19 March 2003).
7. Global Security News, 2 February 2002, www.globalsecurity.org/military/library/news/2001/09/mil-010921-2a613963.htm, (accessed 2 February 2002).
8. "U.S. Military Turns to Video of Sept. 11 to Win the Hearts and Minds of Afghans," *Associated Press*, http://www.whdh.com/news/articles/world/A6392/, (accessed 13 August 2004).
9. Herbert A. Friedman, "Psychological Operations in Afghanistan," http://www.psywarrior.com/Herbafghan.html, (accessed 13 August 2004).
10. Ibid.
11. "Pamphlets Found in Afghanistan Put Price on the Head of Westerners," *Associated Press*, 6 April 2002, http://www.hollandsentinel.com/stories/040602/new_040602002.shtml, (accessed August 13, 2003).
12. "Electronic Warfare: Comprehensive Strategy Still Needed for Suppressing Enemy Air Defenses," General Accounting Office report to the Secretary of Defense. GAO-03-51-Electronic Warfare, 5, November 2002.
13. Taken from report "Operation Iraqi Freedom by the Numbers," USCENTAF, 8, 2003.
14. The new definition of IO in developing joint IO doctrine may not classify this act as IO.
15. "U.S. Takes to the Air(waves) in Afghan Campaign," *Christian Science Monitor* 93, Issue 231, 7, 24 October 2001.
16. Shanker and Schmitt, "Firing Leaflets and Electrons," *New York Times*, 24 February 2003, https://www.us.army.mil/portal/jhtml/earlyBird/Feb2003/e20030224156370.html, (accessed 15 July 2003).
17. Ibid.
18. Douglas Redman and Jack Taylor, "Intelligence and Electronic Warfare Systems Modernization," *Army Magazine*, April 2002, http://www.ausa.org/www/armymag.nsf/0/DCD767B01BDF0AE685256B83006F1BE7?OpenDocument, (accessed 15 July 2003).
19. Taken from a message by Secretary of Defense Donald Rumsfeld, "Web Site OPSEC Discrepancies, Message 141553Z" January 2003, http://www.dod.gov/webmasters/

policy/rumsfeld_memo_to_DOD_webmasters.html on 9 January 2003 (accessed 14 January 2003).

20. Elizabeth Becker, "In the War on Terrorism, a Battle to Shape Opinions," *New York Times*, B4, 11 November 2003.

21. Chris Hogg, "Embedded Journalism and the Rules of War," http://www.digital journal.com/print.htm?id=3586, (accessed 10 July 2003).

22. Paul Sperry, "Marine General Slams 'Chicken Little' News," http://www.worldnet daily.com/news/article.asp?ARTICLE_ID=33378 (accessed 2 July 2003).

23. Bradley Graham, "Bomber Crew Recounts Accounts Attack Operation," *Washington Post*, Section A, 7, 8 April 2003.

24. Andrew Buncombe, "U.S. Army Chief Says Iraqi Troops Took Bribes to Surrender," *The Independent* 24 May 2003, http://news.independent.co.uk/world/middle_east/ story.jsp?story=409090, (accessed 15 July 2003).

25. Kenneth Freed, "U.S. Gave Cedras $1 Million in Exchange for Resignation," *Los Angeles Times*, 14 October 1994, http://www-tech.mit.edu/V114/N48/cedras.48 w.html, (accessed 14 July 2003).

26. Michael Kilian, "U.S. Targets Iraqis' Resolve with Psychological Warfare," *Chicago Tribune*, 3 February 2003,http://www.kansascity.com/mld/kansascity/news/ 5258152.htm?template=contentModules/printstory.jsp (accessed 16 July 2003).

27. Ibid.

28. Defense Science Board, "The Creation and Dissemination of All Forms of Information in Support of Psychological Operations (PSYOP) in Time of Military Conflict," 48, Washington, DC: Office of the Undersecretary of Defense for Acquisition, Technologies, and Logistics , May 2000.

29. "HMS Chatham News," UK Ministry of Defense, 9 April 2003, http://www.royal-navy.mod.uk/static/pages/content.php3?page=5034 (accessed 8 July 2003).

30. John Yurechko, "Iraqi Deception and Denial," U.S. State Department Briefing, 11 October 2002, http://fpc.state.gov/14337.htm, (accessed 4 July 2003). This briefing to the media was intended to show some of the denial and deception techniques the Iraqis had used.

31. "Mobile Labs Found in Iraq," *CNN News*, 15 April 2003, http://us.cnn.com/2003/ WORLD/meast/04/14/sprj.irq.labs/, (accessed 15 April 2003).

32. David Jackson, "U.S. Confident on Weapons," *Dallas Morning News*, 16 April 2003, http://www.bradenton.com/mld/charlotte/news/special_packages/iraq/ 5643233.htm?template=contentModules/printstory.jsp,(accessed on 11 July 2003).

33. Yurechko, "Iraqi Deception and Denial."

34. "Iraq Lessons Learned: Combat Lessons Learned," *Strategy Page*, http://www.strategy page.com/iraqlessonslearned/iraqwarlessonslearned.asp, (accessed 14 July 2003).

35. "Saddam's Surprisingly Friendly Skies," *Business Week Online*, 4 April 2003, http:/ /www.businessweek.com/technology/content/apr2003/tc2003044_3053_ tc119.htm(accessed on 7 July 2003).

36. Ibid.

37. Jim Garamone, "CENTCOM Charts Operation Iraqi Freedom Progress," *American Forces Press Service*, 25 March 2003, http://www.defenselink.mil/news/Mar2003/ n03252003_200303254.html, (accessed 9 July 2003).

38. Shanker and Schmitt, "Firing Leaflets and Electrons," *New York Times*, 23 February 2003, https://www.us.army.mil/portal/jhtml/earlyBird/Feb2003/e20030224156 370.html, (accessed 15 July 2003).

39. "Intercepts: Cyber Battles," *Federal Computer Week*, 7 April 2003, http://www. fcw.com/fcw/articles/2003/0407/intercepts-04-07-03.asp, (accessed 15 April 2003).

40. Dennis Fisher, "Clarke Takes Gov't to Task Over Security," *eWeek*, 15 July 2003, http://www.eweek.com/print_article/0,3668,a=44781,00.asp, (accessed on 8 July 2003).
41. "The General Has His Say," *Parade Magazine*, 1 August 2004, 5.
42. Jo Anne Davis, "Congresswoman Jo Ann Davis Reports Home," an undated, printed newsletter mailed to her constituents in Virginia's 1st Congressional District in August 2004.
43. Ralph Peters, "Kurds' Success Provides Lessons for the Rest of Iraq," *USA Today*, 14 April 2004, http://www.usatoday.com/news/opinion/editorials/2004-04-25-kurds-edit_x.htm, (accessed 24 August 2004).
44. Cheryl Benard, "Afghanistan Without Doctors," *Wall Street Journal*, A10, 12 August 2004.
45. Ollie Ferreira, "Aiding the Enemy," *Daily Press*, A1, 30 July 2004.
46. Faye Bowers, "Terrorists Spread Their Messages Online," *Christian Science Monitor*, 28 July 2004, http://www.csmonitor.com/2004/0728/pp03s01-usgn.html, (accessed 28 July 2004).
47. Gabriel Weimann, "How Modern Terrorism Uses the Internet," United States Institute of Peace, 18 May 2004, http://www.ocnus.net/cgi-bin/exec/view.cgi?archive=46&num=11920, (accessed 20 May 2004).
48. Florian Rötzer, "Terror.net: 'Online-Terorismus' und die Medien," *Telepolis*, 15 July 2004, http://www.telepolis.de/deutsch/special/info/17886/1.html, (accessed 2 July 2004).
49. U.S. Congress, *9/11 Commission Report*, 22 July 2004, 376.
50. Statistics provided by the Broadcasting Board of Governors, http://www.bbg.gov/_bbg_news.cfm?articleID=103&mode=general, (accessed 25 August 2004).
51. "U.S.-Funded Alhurra Television Wins Over Viewers in Iraq," *Space Daily*, 20 July 2004, http://www.spacedaily.com/news/satellite-biz-04zzzze.html, (accessed August 24, 2004).
52. Daoud Kuttab, "America's Clumsy Reach," *Jordan Times*, 2 February 2004, http://www.apfw.org/indexenglish.asp?fname=articles%5Cenglish%5CCare1014.htm (accessed 22 July 2004).
53. Steven A. Cook, "Hearts, Minds, and Hearings," New *York Times*, 6 July 2004, http://www.nytimes.com/2004/07/06/opinion/06COOK.html (accessed 24 August 2004).
54. National Public Radio, "U.S. Says 'Hi' to Young Arabs," 18 August 2003, http://www.npr.org/features/feature.php?wfId=1394868, (accessed 22 August 2004).
55. Islam Online, "U.S. 'Hi' Magazine Eyes Arab 'Future Leaders' Report," 9 August 2003, http://www.islam-online.net/English/News/2003-08/09/article02.shtml, (accessed 23 August 2004).
56. Cynthia Cotts, "Is Anyone Watching the Iraqi Media Network," *Village Voice*, 12–18 November 2003, http://www.villagevoice.com/issues/0346/cotts.php, (accessed 29 June, 2004).
57. Alex Gourevitch, "Exporting Censorship to Iraq," *Vancouver Indymedia*, 11 March 2004, http://www.vancouver.indymedia.org/news/2004/03/114783.php?theme=2 (accessed 25 August 2004).
58. Democracy Now, "U.S. Journalist Quits Pentagon Iraqi Media Project Calling it U.S. Propaganda," 14 January 2004, http://www.democracynow.org/article.pl?sid=04/01/14/1555223, (accessed 23 August 2004).
59. Don North, "One Newsman's Take on How Things Went Wrong," *Zona Europa*, 15 December 2004, http://www.zonaeuropa.com/00945.htm, (accessed on 29 June 2004).
60. "OPSEC Scandal?" *Outside the Beltway*, 8 December 2003, http://www.outsidethebeltway.com/archives/004136.html, 8 July 2004.

61. "U.S. Forces Order of Battle," *Global Security*, 21 May 2004, http://www.global security.org/military/ops/iraq_orbat_040521.htm, (accessed 23 May 2004).

62. From the 2004 National OPSEC Conference agenda, www.nacic.gov/events/doc/ IOSS_Conference_abstracts.pdf, (accessed 8 July 2004).

63. Bryan D. Brown testimony to the Senate Armed Services Committee, 29 July 2003, http://www.infowar-monitor.net/print.php?sid=499, (accessed 8 July 2004).

64. Federal Business Opportunities, Announcement no. 0885, 29 April 2004, http:// www.fbodaily.com/archive/2004/04-April/29-Apr-2004/FBO-00573918.htm, (accessed 8 July 2004).

65. "Operation Enduring Freedom—One Year of Accomplishments," USEUCOM Public Affairs Office, 7 October 2002, http://www.eucom.mil/Directorates/ECPA/ index.htm?http://www.eucom.mil/Directorates/ECPA/Operations/oef/ operation_enduring_freedomII.htm&2 (accessed 15 July 2003).

66. "NATO Expands Presence in Afghanistan," *Relief Web*, 29 June 2004, http:// www.reliefweb.int/w/rwb.nsf/480fa8736b88bbc3c12564f6004c8ad5/ a27413945e61364b85256ec200591b87?OpenDocument, (accessed 5 July 2004).

67. Jim Wagner, "Civil Affairs Soldiers Adapt to New Mission," *Defend America*, February 2003, http://www.defendamerica.mil/articles/feb2003/a010303a.html, (accessed 5 July 2004).

68. Donna Miles, "Terrorists Can't Compete with Provisional Reconstruction Teams," *American Forces Press Service*, 21 April 2004, www.freerepublic.com/focus/news/ 1121837/posts, (accessed 5 July 2004).

69. Richard C. Sater, "Provisional Reconstruction Team Begins Work in Heart," *Army News Service*, 12 April 2003, http://www.freerepublic.com/focus/f-news/1034522/ posts, (accessed 29 July 2004).

70. Miles, "Terrorists Can't Compete."

71. "Civil Affairs Soldiers Rebuild Iraq," *Associated Press*, 13 July 2004, http:// www.military.com/NewsContent/0,13319,FL_rebuild_071304,00.html, (accessed 24 August 2004).

72. Paul Bremer, "Transcript: Paul Bremer on 'Fox News Sunday,'" *Fox News*, 4 July 2004, http://www.foxnews.com/story/0,2933,124686,00.html, (accessed 5 July 2004).

73. James Careless, "Combat Camera Technology Shares Footage Faster from the Field," Government Video, 26 May 2004, http://governmentvideo.com/articles/publish/ article_419.shmtl, (accessed August 25, 2004).

74. Gen. Tommy Franks, Statement before the Senate Armed Services Committee, 7 February 2002.

75. J. Michael Waller, "Losing a Battle for Hearts and Minds," *Insight Magazine*, 1 April 2002, http://www.insightmag.com/main.cfm/include/detail/storyid/225520.html (accessed on 9 April 2002).

76. Margaret Belknap, "The CNN Effect: Strategic Enabler or Operational Risk?" *Parameters*, Autumn 2002, 110.

77. Waller, "Losing a Battle for Hearts and Minds."

78. Donald Rumsfeld, DoD Press Conference, 18 November 2002, http:// www.defenselink.mil/transcripts/2002/t11212002_t1118sd2.html, (accessed 24 July 2004).

79. Jim Garamone, "General Myers Speaks About the Importance of Focused National Power" *Armed Forces News Service*, 16 November 2001, http://www.af.mil/news/ Nov2001/n20011116_1644.shtml (accessed 13 August 2003).

80. V. Adm. Thomas R. Wilson, Testimony Before the Senate Select Committee on Intelligence, 6 February 2002.

81. Garamone, "General Myers Speaks."

82. Todd Zwillich, "Americans Unprepared for Psychological Terror," *Web MD Health*, 19 June 2004, http://my.webmd.com/content/article/89/100345.htm, (accessed 5 July2004).

83. Aired on a live broadcast of *CNN Headline News* at 11:06 AM, Eastern daylight savings time, 7 July 2004.

Chapter 5: Cyberterrorism: Hype and Reality

1. Walter Laqueur, *The New Terrorism: Fanaticism and the Arms of Mass Destruction* (Oxford: Oxford University Press, 1999), 254.

2. John Deutch, statement before the U.S. Senate Governmental Affairs Committee (Permanent Subcommittee on Investigations), 25 June 1996, http://www.nswc.navy.mil/ISSEC/Docs/Ref/InTheNews/fullciatext.html (accessed 25 May 2006).

3. Ralf Bendrath, "The American Cyber-Angst and the Real World: Any Link?" in Robert Latham ed., *Bombs and Bandwidth: The Emerging Relationship Between Information Technology and Security* (New York: New Press, 2003), 49.

4. Center for Strategic and International Studies (CSIS), *Cybercrime, Cyberterrorism, Cyberwarfare: Averting an Electronic Waterloo* (Washington, DC: CSIS Press, 1998), xiii.

5. Barry C. Collin, "The Future of Cyberterrorism," paper presented at the Eleventh Annual International Symposium on Criminal Justice Issues, University of Illinois at Chicago, 1997, http://afgen.com/terrorism1.html (accessed 25 May 2006); Matthew G. Devost, Brian K. Houghton, and Neal Allen Pollard, "Information Terrorism: Political Violence in the Information Age," *Terrorism and Political Violence* 9:1 (1997): 72–83; Mark M. Pollitt, "Cyberterrorism: Fact or Fancy?" *Computer Fraud and Security* (February 1998): 8–10.

6. Bendrath, "The American Cyber-Angst and the Real World" 51–52.

7. Ibid., 51.

8. Ibid., 52.

9. Dorothy Denning, "Is Cyber Terror Next?" in *Understanding September 11*, ed., Craig Calhoun, Paul Price, and Ashley Timmer (New York: New Press, 2001); full text found at http://www.ssrc.org/sept11/essays/denning.htm (accessed 25 May 2006); Jon Swartz, "Experts: Cyberspace Could Be Next Target," *USA Today*, 16 October 2001.

10. Francis Richardson, "Cyberterrorist Must Serve Year in Jail," *Boston Herald*, 6 June 2001.

11. Still, Kathy, "Wise County Circuit Court's Webcam 'Cracked' by Cyberterrorists," *Bristol Herald Courier*, 20 December 2001.

12. Institute for Security Technology Studies (ISTS), *Cyber Attacks During the War on Terrorism: A Predictive Analysis* (Dartmouth College: Institute for Security Technology Studies, 2001). Full text available online at http://www.ists.dartmouth.edu/analysis/cyber_a1.pdf (accessed 25 May 2006).

13. linkLINE Communications, Inc., "linkLINE Communications Thwarts Cyber-Terrorist," Yahoo!Finance, 19 March 2002.

14. John Schwartz, "When Point and Shoot Becomes Point and Click,' *New York Times,* 12 November 2000.

15. Collin, "The Future of Cyberterrorism."

16. As quoted in Devost, Houghton, and Pollard, "Information Terrorism: Political Violence in the Information Age," 76.

17. National Research Council, *Computers at Risk: Safe Computing in the Information Age* (Washington DC: National Academy Press, 1991), 7. Full text found at http://www.nap.edu/books/0309043883/html/index.html (accessed 25 May 2006).

18. Sarah Gordon and Richard Ford, "Cyberterrorism?" *Computers and Security* 21:7 (2002): 636.

19. Ayn Embar-Seddon, "Cyberterrorism: Are We Under Siege?" *American Behavioral Scientist* 45:6 (2002): 1034.

20. Barry Collin, quoted in James D. Ballard, Joseph G. Hornik, and Douglas McKenzie, "Technological Facilitation of Terrorism: Definitional, Legal, and Policy Issues," *American Behavioral Scientist* 45:6 (2002): 992.

21. Collin, "The Future of Cyberterrorism."

22. William Gibson, *Neuromancer* (New York: Ace, 2004).

23. See, for example, Conor Gearty, Terror (London: Faber & Faber, 1998); Adrian Guelke, *The Age of Terrorism and the International Political System* (London and New York: I. B. Tauris Publishers, 1998); Bruce Hoffman, *Inside Terrorism* (London: Indigo, 1998); Alex P. Schmid and Albert J. Jongman, Political Terrorism: A New Guide to Actors, Authors, Concepts, Databases, Theories and Literature (Amsterdam: North-Holland Publishing, 1988); Grant Wardlaw, *Political Terrorism: Theory, Tactics, and Countermeasures* (Cambridge: Cambridge University Press, 1982).

24. Pollitt, "Cyberterrorism: Fact or Fancy?" 9.

25. Denning, "Is Cyber Terror Next?"; also contained in Denning's testimony before the Special Oversight Panel on Terrorism, Committee on Armed Services, U.S. House of Representatives, 23 May 2000, http://www.cs.georgetown.edu/~denning/infosec/cyberterror.html (accessed 25 May 2006). Denning first put forward this definition in her 1999 paper "Activism, Hacktivism, and Cyberterrorism: The Internet as a Tool for Influencing Foreign Policy," http://www.rand.org/publications/MR/MR1382/MR1382.ch8.pdf.

26. Gordon and Ford, "Cyberterrorism?" 637.

27. See, among others, CSIS, *Cybercrime, Cyberterrorism, Cyberwarfare*; Christen, Denney and Maniscalco, "Weapons of Mass Effect: Cyber-Terrorism"; Mark Henych, Stephen Holmes, and Charles Mesloh, "Cyber Terrorism: An Examination of the Critical Issues," *Journal of Information Warfare* 2:2 (2003); Rattray, "The Cyberterrorism Threat."

28. Gordon and Ford, "Cyberterrorism?" 640.

29. Devost, Houghton, and Pollard, "Information Terrorism: Political Violence in the Information Age," 75.

30. Ibid., 76.

31. Guelke, *The Age of Terrorism and the International Political System,* 19; Michael Mates, Technology and Terrorism (Brussels: NATO, 2001), http://www.tbmm.gov.tr/natopa/raporlar/bilim%20ve%20teknoloji/AU%20121%20STC%20Terrorism.htm (accessed 25 May 2006); Schmid and Jongman, *Political Terrorism: A New Guide,* 5.

32. Devost, Houghton, andPollard, "Information Terrorism: Political Violence in the Information Age," 10.

33. Bill Nelson, Rodney Choi, Michael Iacobucci, Mark Mitchell, and Greg Gagnon, *Cyberterror: Prospects and Implications* (Monterey, CA: Center for the Study of Terrorism and Irregular Warfare, 1999), 7. Full text available at http://www.nps.navy.mil/ctiw/files/Cyberterror%20Prospects%20and%20Implications.pdf(accessed 25 May 2006).

34. Ibid., 8.

35. As quoted in Nelson, et al., *Cyberterror: Prospects and Implications*, 9.
36. Ibid., 9.
37. H. Sher, "Cyberterror Should be International Crime—Israeli Minister," *Newsbytes* 10 November 2000.
38. Sankei Shimbun, "Government Sets Up Anti-Cyberterrorism Homepage," Foreign Broadcast Information Service (FBIS), (FBIS-EAS-2002-0410), 10 April 2002.
39. "Russia Cracks Down on 'Cyberterrorism,'" Foreign Broadcast Information Service (FBIS), ITAR-TASS (FBIS-SOV-2002-0208), 8 February, 2002.
40. Tanya Hershman, "Cyberterrorism is Real Threat, Say Experts at Conference," 11 December 2000, israel.internet.com.
41. See National Communications System, "The Electronic Intrusion Threat to National Security and Emergency Preparedness (NS/EP) Internet Communications: An Awareness Document" (Arlington, VA: Office of the Manager, National Communications Systems, 2000), 40. Full text available at http://www.ncs.gov/library/reports/ electronic_intrusion_threat2000_final2.pdf (accessed 25 May 2006).
42. As quoted in Liz Duff and Simon Gardiner, "Computer Crime in the Global Village: Strategies for Control and Regulation—In Defense of the Hacker," *International Journal of the Sociology of Law* 24:2 (1996): 215.
43. As quoted in Amanda Chandler, "The Changing Definition and Image of Hackers in Popular Discourse," *International Journal of the Sociology of Law* 24:2 (1996): 232.
44. Kevin Soo Hoo, Seymour Goodman, and Lawrence Greenberg, "Information Technology and the Terrorist Threat," *Survival* 39:3 (1997): 144–145; Gregory J. Rattray, "The Cyberterrorism Threat,'" in *The Terrorism Threat and U.S. Government Response: Operational and Organizational Factors*, ed., James M. Smith and William C. Thomas (Colorado: U.S. Air Force Institute for National Security Studies, 2001), 89. The full text of the latter is available at http://www.usafa.af.mil/df/inss/Ch%205.pdf (accessed 25 May 2006).
45. See, for example, Chandler, "The Changing Definition and Image of Hackers in Popular Discourse," 242–246; Duff and Gardiner, "Computer Crime in the Global Village," 223; Reid Skibell, "The Myth of the Computer Hacker," *Information, Communication and Society* 5:3 (2002): 342; Paul A. Taylor, *Hackers: Crime in the Digital Sublime* (London: Routledge, 1999), 44–50.
46. CSIS, *Cybercrime, Cyberterrorism, Cyberwarfare*, 15.
47. See Clifford Stoll, *The Cuckoo's Egg* (London: Pan Books, 1991).
48. See Jack L. Brock, *Computer Security: Hackers Penetrate DoD Computer Systems* (Washington, DC: General Accounting Office, 1991). Full text available online at http://www.globalsecurity.org/security/library/report/gao/145327.pdf (accessed 25 May 2006).
49. See Andrew Rathmell, Richard Overill, Lorenzo Valeri, and John Gearson, "The IW Threat from Sub-State Groups: An Interdisciplinary Approach," paper presented at the Third International Symposium on Command and Control Research and Technology, Institute for National Strategic Studies, National Defense University, Washington, DC, 17–20 June, 1997, 4. Full text available at http://www.kcl.ac.uk/orgs/ icsa/Old/terrori.html (accessed 25 May 2006).
50. Rattray, "The Cyberterrorism Threat," 87–88.
51. Rathmell, et al., "The IW Threat from Sub-State Groups," 5.
52. David Tucker, "The Future of Armed Resistance: Cyberterror? Mass Destruction?" Conference Report (Monterey, CA: The Center on Terrorism and Irregular Warfare, 2000), 16. Full text available at http://www.nps.navy.mil/ctiw/files/ substate_conflict_dynamics.pdf (accessed 25 May 2006).

53. Ibid., 14–16.
54. Embar-Seddon, "Cyberterrorism: Are We Under Siege?" 1037.
55. Soo Hoo, Goodman, and Greenberg, "Information Technology and the Terrorist Threat," 141.
56. John Borland, "Analyzing the Threat of Cyberterrorism," *TechWeb: The Business Technology Network*, 25 September 1998, http://www.techweb.com/wire/story/ TWB19980923S0016 (accessed 25 May 2006). See also Andrew Rathmell, "Cyber-Terrorism: The Shape of Future Conflict?" RUSI Journal, October (1997): 43–44, at http://www.kcl.ac.uk/orgs/icsa/Old/rusi.html (accessed 25 May 2006); Rathmell, et al., "The IW Threat from Sub-State Groups: An Interdisciplinary Approach," 7–8.
57. Rattray, "The Cyberterrorism Threat," 89; Jessica Stern, *The Ultimate Terrorists* (Cambridge, MA: Harvard University Press, 1999), 74; Lorenzo Valeri and Michael Knights, "Affecting Trust: Terrorism, Internet and Offensive information Warfare," *Terrorism and Political Violence* 12:1 (2000): 20.
58. Borland, "Analyzing the Threat of Cyberterrorism."
59. Kevin O'Brien and Joseph Nusbaum, "Intelligence Gathering on Asymmetric Threats: Part 2," *Jane's Intelligence Review* 15:11 (2000): 53.
60. See Martha Mendoza, "Virus Sender Helped FBI Bust Hackers, Court Records Say," *USA Today*, 18 September 2003, http://www.usatoday.com/tech/news/computer security/2003-09-18-reformed-hacker_x.htm (accessed 25 May 2006).
61. Soo Hoo, Goodman, and Greenberg, "Information Technology and the Terrorist Threat," 143.
62. Kevin O'Brien and Joseph Nusbaum, "Intelligence Gathering on Asymmetric Threats: Part 1," *Jane's Intelligence Review* 15:10 (2000): 53.
63. Hank T. Christen, James P. Denney, and Paul M. Maniscalco, "Weapons of Mass Effect: Cyber-Terrorism," in Paul M. Maniscalco and Hank T. Christen, eds., *Understanding Terrorism and Managing the Consequences* (Upper Saddle River, NJ: Prentice Hall, 2002), 194.
64. As quoted in Skibell, "The Myth of the Computer Hacker," 342.
65. Soo Hoo, Goodman, and Greenberg, "Information Technology and the Terrorist Threat," 145.
66. See Amy Harmon, "'Hacktivists' of All Persuasions Take Their Struggle to the Web," *New York Times*, 31 October 1998, http://www.cs.du.edu/~lavita/hacktivists.pdf (accessed 25 May 2006); Niall McKay, "The Golden Age of Hacktivism," *Wired*, 22 September 1998, http://www.wirednews.com/news/politics/0,1283,15129,00.html (accessed 25 May 2006). See also Douglas Thomas, "Finding a New Term: From 'Hacking' to 'Cybercrime,'" *Online Journalism Review*, 22 February 2000, http:// www.ojr.org/ojr/ethics/1017965933.php (accessed 25 May 2006).
67. McKay, "The Golden Age of Hacktivism"; Alexandra Samuel, "Digital Disobedience: Hacktivism in Political Context," paper presented at the American Political Science Association (APSA) Annual Conference, San Francisco, CA, 29 September –2 August 2001; Stefan Wray, "Electronic Civil Disobedience and the World Wide Web of Hacktivism: A Mapping of Extraparliamentarian Direct Action Net Politics," paper presented at The World Wide Web and Contemporary Cultural Theory Conference, Drake University, November 1998, http://switch.sjsu.edu/web/v4n2/stefan/.
68. Dorothy Denning, "Cyberwarriors: Activists and Terrorists Turn to Cyberspace," *Harvard International Review* 23:2 (2001), http://www.hir.harvard.edu/articles/ index.html?id=905 (accessed 25 May 2006); Denning, "Activism, Hacktivism, and Cyberterrorism," 25–26.
69. Denning, "Cyberwarriors: Activists and Terrorists Turn to Cyberspace."

70. See Harmon, "'Hacktivists' of All Persuasions Take Their Struggle to the Web" and McKay, "The Golden Age of Hacktivism."

71. Michael Vatis, "What is Cyber-Terrorism?" in Yonah Alexander and Michael S. Swetnam, eds., *Cyber Terrorism and Information Warfare: Threats and Responses* (New York: Transnational Publishers, 2001), 4.

72. Tim Jordan, "Mapping Hacktivism: Mass Virtual Direct Action (MVDA), Individual Virtual Direct Action (IVDA) and Cyberwars," *Computer Fraud & Security* 4 (2001): 8.

73. Ibid., 9.

74. Wray, "Electronic Civil Disobedience and the World Wide Web of Hacktivism," 7.

75. See Jordan, "Mapping Hacktivism," 11; and Wray, "Electronic Civil Disobedience and the World Wide Web of Hacktivism," 11.

76. Wray, "Electronic Civil Disobedience and the World Wide Web of Hacktivism," 3. See also Denning, "Activism, Hacktivism, and Cyberterrorism"; and Jordan, "Mapping Hacktivism," 10.

77. Samuel, "Digital Disobedience," 4.

78. As quoted in Harmon, "'Hacktivists' of All Persuasions Take Their Struggle to the Web."

79. Ronald Deibert, *Black Code: Censorship, Surveillance, and the Militarization of Cyberspace* (New York: Social Science Research Council, 2003), 19n64, http://www.ssrc.org/programs/itic/publications/ITST_materials/blackcode.pdf (accessed 25 May 2006); Denning, "Activism, Hacktivism, and Cyberterrorism."

80. See Denning's testimony before the Special Oversight Panel on Terrorism.

81. Mates, *Technology and Terrorism.*

82. Thomas, "Finding a New Term."

83. See National Communications System, The Electronic Intrusion Threat to National Security and Emergency Preparedness (NS/EP) Internet Communications, 36–39.

84. CSIS, *Cybercrime, Cyberterrorism, Cyberwarfare*, 3.

85. National Communications System, The Electronic Intrusion Threat to National Security and Emergency Preparedness (NS/EP) Internet Communications, 35.

86. Gordon and Ford, "Cyberterrorism?" 636—637, 641.

87. Ibid., 637.

88. Mates, *Technology and Terrorism*, 6.

89. Nelson, et al., *Cyberterror: Prospects and Implications*, 9–10.

90. Nelson, et al., *Cyberterror: Prospects and Implications*, 10. See also Linda Garrison and Martin Grand, "Cyberterrorism: An Evolving Concept," National Infrastructure Protection Center: Highlights 6:01 (2001): 3, http://www.iwar.org.uk/infocon/nipc-highlights/2001/highlight-01-06.pdf (accessed 25 May 2006).

91. Kirsten Weisenberger, "Hacktivists of the World, Divide," SecurityWatch.com, (accessed 23 April 2001).

92. See, for example, Maura Conway, "Terrorism and the Internet: New Media, New Threat?" *Parliamentary Affairs* 59:2 (2006); Maura Conway, "Cybercortical Warfare: Hizbollah's Internet Strategy," in Sarah Oates, Diana Owen, and Rachel Gibson, eds., *The Internet and Politics: Citizens, Voters and Activists* (London: Routledge, 2005); Maura Conway, "Terrorist Web Sites: Their Contents, Functioning, and Effectiveness," in Philip Seib, ed., *Media and Conflict in the Twenty-First Century* (New York: Palgrave, 2005); Maura Conway, "Reality Bytes: Cyberterrorism and Terrorist 'Use' of the Internet," First Monday 7:11 (2002), http://www.firstmonday.org/issues/issue7_11/conway/index.html (accessed 25 May 2006); Gabriel Weimann, *Terror on the Internet: The New Arena, the New Challenges* (Washington, DC: United

States Institute of Peace Press, 2006); Gabriel Weimann, "www.terror.net: How Modern Terrorism Uses the Internet" (Washington DC: United States Institute of Peace, 2004), http://www.usip.org/pubs/specialreports/sr116.pdf (accessed 25 May 2006).

93. See Paola Di Maio, "Hacktivism, Cyberterrorism or Online Democracy?' Information Warfare Site, 2001, http://www.iwar.org.uk/hackers/resources/hacktivism-europe/internet-europe.htm (accessed 25 May 2006); also Mates, *Technology and Terrorism*.

94. Kirsten Weisenberger, "Hacktivists of the World, Divide," 9.

95. NIPC, 2001;Middleton, 2002; Levin, 2002, 984–985.

96. As quoted in Declan McCullagh, "Bush Signs Homeland Security Bill," ZDNet, 25 November 2002, http://news.zdnet.com/2100-1009_22-975305.html (accessed 25 May 2006).

97. Kevin Poulsen, "Lawyers Fear Misuse of Cyber Murder Law," *Security Focus Online*, 21 November 2001, http://online.securityfocus.com/news/1702 (accessed 25 May 2006); McCullagh, "Bush Signs Homeland Security Bill."

98. Denning, "Cyberwarriors."

99. Carole Veltman, "Beating Cyber Crime," *Daily Telegraph*, 1 March 2001: 12E.

100. Poulsen, "Lawyers Fear Misuse of Cyber Murder Law."

Chapter 6: Information Operations Education: Lessons Learned from Information Assurance

1. Robert Keohane and Joseph Nye, *Power and Interdependence*, (Boston: Longman, 1989), 23.

2. John Arquilla and David Ronfeldt, "Looking Ahead: Preparing for Information-Age Conflict," in *Athena's Camp: Preparing for Conflict in the Information Age*, ed., John Arquilla and Daid Ronfeldt (Santa Monica, CA: Rand, 1997) 441.

3. The first known use of information warfare was in a briefing title and concept written by Dr. Tom Rona (then of Boeing) for Andrew Marshall, May/June 1976.

4. The first document, Department of Defense Document (DoDD) TS3600.1, was kept at the Top Secret level throughout its use, due to the restrictive nature of this new strategy.

5. This author (Dr. Kuehl) vividly remembers the very first group of students at the School of Information Warfare and Strategy and the dismay with which they greeted the plethora of IW definitions during the initial class session in August 1995. Table 1 – IO Training and Education Capabilities, located at http://cryptome.org/iwd.htm, November 1, 2006.

6. Corey Schou, W. V. Maconachy, and James Frost, "Organizational Information Security: Awareness, Training and Education to Maintain System Integrity," in Proceedings Ninth International Computer Security Symposium, Toronto, Canada, May 1993.

7. Accessed at http://www.globalsecurity.org/intell/library/congress/1997_hr/h970211s.htm.

8. Estimated from the Taulbee Survey conducted by the Computing Research Association; see http://cra.org.

9. Ê.Øíó, Ä.Ôðèíê, Ä.Äÿâèñ, Ä.Ôðîñò. Îáó÷åíèå èíôîðìàöèîííîé áåçïàñíîñòè: ñîâûèãîè êðèçèñ è ïîäàîòîâêà ñïåöèàëèñòîâ // Áåçïàñíîñòü èíôîðìàöèîííûõ òåõíîëîãèé. *[Training of information safety: the world crisis and preparation of experts, Safety of information technologies]*, 4. - 2001. - Ñ. 5-10. *(In Russian); Corey D. Schou, Deborah*

Frincke, James Davis, and James Frost, Information Security Education: A Worldwide Workforce Crisis, *WISE 2 Proceedings, IFIP 11.8, Perth, Australia, June 2001.*

Chapter 7: Information Operations and the Average Citizen

1. Leigh Armistead, ed., *Information Operations: Warfare and the Hard Reality of Soft Power,* (Washington, DC: Brassey's, 2004), 49.
2. Gregg Keizer, "Unprotected PCs Fall to Hacker Bots in Just Four Minutes," 30 November 2004, http://www.techweb.com/wire/security/54201306.
3. Seymour Bosworth and M. E. Kabay, eds., *Computer Security Handbook,* 4th ed. (New York: Wiley & Sons, 2002), 9:18.
4. Christine Tatum, "Legendary Con Artist Warns: Hackers Play Mind Games," *Chicago Tribune,* 26 June 2003.
5. http://www.quotes.stevenredhead.com/Sun-tzu.html.
6. Erik Sherman, "Foreign Policy," *Information Security Magazine,* September 2004, 56.
7. http://www.zone-h.com/en/defeacements.
8. Vympel, Zone-H Admin, "Brazilian Defacers Hack Hundreds of Stanford University Web Sites," 21 August 2005, http://www.zone-h.org/en/news/read/id=205962.
9. Grant Gross, "Experts: Worry More About Insiders Than Cyberterrorism," *IDG News Service,* 3 June 2003, http://www.nwfusion.com/news/2003/0603terrorism.html.
10. http://searchsecurity.techtarget.com/gDefinition/0,294236,sid14_gci214518,00.html.
11. Shawna McAlearny, "Experts Predict New Path for Malicious Code, Antivirus Products," 16 June 2005, http://searchsecurity.techtarget.com/originalContent/0,289142,sid14_gci1098714,00.html.
12. Robert McMillan, "Exploits on the Loose for Latest Microsoft Bugs," *IDG News Service,* 12 August 2005, http://www.networkworld.com/news/2005/08/1205-microsoft-bugs.html.
13. http://www.staysafeonline.info/home-glossary.html.
14. Bill Brenner, "New Malcode Disguised as Fake Security," CNN bulletins, 21 January 2005, http://searchsecurity.techtarget.com/originalContent/0,289142,sid14_gci1046879,00.html?track=NL-102&ad=501293.
15. Jeff Crume, *Inside Internet Security: What Hackers Don't Want You to Know* (New York, Addison-Wesley), 2000.
16. SearchSecurity.com staff, "Security Bytes: Worm Uses Spycam," 24 August 2004, http://searchsecurity.techtarget.com/originalContent/0,289142,sid14_gci1001983,00.html?track=NL-102&ad=489996.
17. Bill Brenner, "Malware Roundup: The Fast and Furious of 2004," 3 January 2005, http://searchsecurity.techtarget.com/originalContent/0,289142,sid14_gci1037997,00.html.
18. Gregg Keizer, "Unprotected PCs Fall to Hacker Bots in Just Four Minutes," 30 November 2004, http://www.techweb.com/wire/security/54201306.
19. Paul F. Roberts, "Worms Exploit Plug and Play Vulnerabilities," *eWeek,* 22 August 2005.
20. Microsoft Corp., "WD: How Word for Windows Uses Temporary Files," Microsoft Knowledge Base Article, 211632. http://support.microsoft.com/default.aspx?scid=http://support.microsoft.com:80/support/kb/articles/ Q211/6/32.ASP&NoWebContent=1.
21. Sandy Berger, "How to Eliminate Temp Files," *AARP How to Guides,* http://www.aarp.org/computers-howto Sandy /Articles/a2002-07-15-tempfiles.html.

22. Richard M. Smith, "Microsoft Word Bytes Tony Blair in the Butt," 30 June 2003, http://www.computerbytesman.com/privacy/blair.htm.
23. Lenny Bailes, "Seek Out Hidden Files," *PC World*, 8 August 2000, http://www.pcworld.com/howto/article/0,aid,12834,00.asp.
24. Hewlett-Packard, "Snoop Proof Your PC," http://h71036.www7.hp.com/hho/cache/836-0-0-225-121.aspx.
25. Scott Spanbauer, "Internet Tips: Protect Yourself—Clear Your Cookies and History," PC World Magazine, *January 2003, http://www.pcworld.com/howto/article/0,aid,106715,00.asp.*
26. "Net Threat Rising," *Consumer Reports*, September 2005, 12.
27. Ibid., 13.
28. Ibid., 14.
29. Vilis Ostis, "Gateways Are the Best Way to Stop Spyware," Network World, 20 June 2005, http://www.networkworld.com/columnists/2005/062005-spyware-gateway.html.

Chapter 8: A Tale of Two Cities: Approaches to Counterterrorism and Critical Infrastructure

1. James A. Lewis, *Assessing the Risks of Cyber Terrorism, Cyber War and Other Cyber Threats,* (Washington, DC: Center for Strategic and International Studies, 2002); Maura Conway, "Cyberterrorism: The Story So Far," *Journal of Information Warfare* 2:2 (February 2003): 33–42; Mathew Devost, "Cyberterrorism: Identifying and Responding to New Threats," Presentation to the Information Operations Europe 2003 Conference, London, 2–3 July 2003.
2. Thomas Homer-Dixon, "The Rise of Complex Terrorism," *Foreign Policy*, January–February 2002: 52–62.
3. Richard L. Millet, "Vague Threats and Concrete Dangers: The Global Security Environment at the Start of the Twenty-first Century," in Max G. Manwaring, et al., eds., *The Search for Security: A U.S. Grand Strategy for the Twenty-First Century,* (Westport, CT: Praeger, 2003).
4. Executive Office of the President, Presidential Decision Directive (PDD) 39 *U.S. Counterterrorism Policy,* (Washington, DC: Executive Office of the President, 21 June 1995). The document was originally classified Secret, but was declassified with some deletions on 24 January 1999, http://www.fas.org/irp/offdocs/pdd39.htm (accessed 2 June 2006).
5. Ibid.
6. Anthony H. Cordesman and Justin G. Cordesman, *Cyber-threats, Information Warfare and Critical Infrastructure Protection: Defending the U.S. Homeland,* (Westport, Connecticut: Praeger, 2002): 1–2.
7. Ibid.
8. PDD 39; Michael Vatis, Statement for the Record on Infrastructure Protection and the Role of the National Infrastructure Protection Center, before the Senate Judiciary Sub-Committee on Technology, Terrorism, and Government Information, Washington, DC, 10 June 1998.
9. Ibid.
10. Executive Office of the President, Executive Order 13010, *Critical Infrastructure Protection,* (Washington, DC: Executive Office of the President, 17 July 1996), http://frwebgate.access.gpo.gov/cgi-bin/getdoc.cgi?dbname=1996_register&docid=fr17jy96-92.pdf (accessed 2 June 2006).

11. National Communications System, "The Electronic Intrusion Threat to National Security and Emergency Preparedness Telecommunications," (Washington, DC: National Communications System, 1994), http://handle.dtic.mil/100.2/ADA301642 (accessed 2 June 2006); Information Infrastructure Task Force, "NII Security: The Federal Role, Draft Report," (Washington, DC: Department of Commerce), 5 June 1995), http://nsi.org/Library/Compsec/nii.txt (accessed 2 June 2006); Science Applications International Corporation (SAIC), "Information Warfare: Legal, Regulatory, Policy and Organizational Considerations for Assurance," (McLean, VA: SAIC, July 1995), http://handle.dtic.mil/100.2/ADA316285 (accessed 2 June 2006); Defense Science Board (DSB), "Report of the DSB on Information Warfare-Defense", (Washington, DC: DSB, November 1996) http://www.acq.osd.mil/dsb/reports/iwd.pdf (accessed 2 June 2006); Commission on Protecting and Reducing Government Secrecy, "Report of the Commission on Protecting and Reducing Government Secrecy," (Washington, DC: U.S. Government Printing Office, 1997), http://www.gpo.gov/congress/commissions/secrecy/index.html (accessed 2 June 2006); Gregory Rattray, *Strategic Warfare in Cyberspace*, (Cambridge, MA.: MIT Press, 2001), 339–40.

12. Vatis, "Statement for the Record on Infrastructure Protection and the Role of the National Infrastructure Protection Center."

13. Ibid.

14. Ibid.

15. Presidential Commission on Critical Infrastructure Protection (PCCIP), "Critical Foundations: Protecting America's Infrastructures," (Washington, DC: PCCIP, October 1997), http://www.fas.org/sgp/library/pccip.pdf (accessed 2 June 2006).

16. Vatis, "Statement for the record on Infrastructure Protection and the Role of the National Infrastructure Protection Center."

17. Executive Office of the President, Presidential Decision Directive (PDD) 62, *Protection Against Unconventional Threats to the Homeland and Americans Overseas,* (Washington, DC: Executive Office of the President, 22 May 1998), http://www.fas.org/irp/offdocs/pdd-62.htm (accessed 2 June 2006); Executive Office of the President, Presidential Decision Directive (PDD) 63, Critical Infrastructure Protection, (Washington, DC: Executive Office of the President, 22 May 1998), http://www.fas.org/irp/offdocs/pdd/pdd-63.htm (accessed 2 June 2006).

18. Ibid.

19. Ibid.

20. Ibid.

21. Executive Office of the President, *Defending America's Cyberspace—National Plan for Information Systems Protection 1.0— An Invitation to Dialogue,* (Washington, DC: Executive Office of the President, January 2000), http://www.fas.org/irp/offdocs/pdd/CIP-plan.pdf (accessed 2 June 2006).

22. John D. Moteff, "Critical Infrastructures: Background, Policy and Implementation," (Washington, DC: Congressional Research Service, Report No. RL30153, 7 August 2003): 19, http://www.law.umaryland.edu/marshall/crsreports/crsdocuments/RL30153_08072003.pdf (accessed 2 June 2006).

23. Ibid.

24. Ibid.

25. Ibid.

26. Paul Dibb, *The Conceptual Basis of Australia's Defence Planning and Force Structure Development*, (Canberra: Strategic and Defence Studies Centre, Australian National University, Canberra Paper No. 88, March 1992) 1–15.

27. Jeff Malone, "Low-Intensity Conflict and Australian National Security Policy,"

(unpublished Master of Arts thesis, Department of Political Science, The University of Western Australia, 1997), 25.

28. Robert M. Hope, "Report of the Protective Security Review—Unclassified Version," (Canberra: Australian Government Publishing Service, May 1979).

29. Ibid, 99.

30. Standing Advisory Committee on Commonwealth/State Cooperation for Protection Against Violence (SAC-PAV), "Report of the 1993 SAC-PAV Review," (Canberra: SAC-PAV, January 1994): i.

31. Jeff Malone, "Low-Intensity Conflict and Australian National Security Policy," 22–38.

32. J. A. Sheldrick, "The Vital Installations Program," in John O. Langtry and Des Ball, eds., *A Vulnerable Country? Civil Resources in the Defence of Australia*, (Canberra: Strategic and Defence Studies Centre, Australian National University, 1986), 512–520.

33. Michael H. Codd, "Review of Plans and Arrangements in Relation to Counter-Terrorism," (Canberra: Department of Prime Minister and Cabinet, 25 May 1992); SAC-PAV, "Report of the 1993 SAC-PAV Review"; i.

34. Inter-Departmental Committee on Protection of the National Information Infrastructure (IDC-PNII), "Report of the Inter-Departmental Committee on Protection of the National Information Infrastructure," (Canberra: Attorney-General's Department, December 1998), 1.

35. Ibid.

36. Daryl Williams, "Protecting Australia's Information Infrastructure," Media Release, 26 August 1999, http://www.ag.gov.au/agd/WWW/attorneygeneralHome.nsf/Page/Media_Releases_1999_August_Protecting_Australia's_Information_Infrastructure (accessed 2 June 2006).

37. Geoffrey Barker, "First Move to Beat Cyber Attacks on Vital Systems," *Australian Financial Review*, 30 August 1999, 6.

38. Clive Williams, "The Sydney Olympics: The Trouble-Free Games," (Canberra: Strategic and Defence Studies Centre, Australian National University, Working Paper No. 371, November 2002), 6–8.

39. Ibid., 8–9.

40. Athol Yates, "Engineering a Safer Australia: Securing Critical Infrastructure and the Built Environment," (Canberra: Institute of Engineers Australia, June 2003), 8–9, http://www.homelandsecurity.org.au/files/EngineeringaSaferAust.pdf (accessed 2 June 2006).

41. Richard Crothers, et al., "The AFP Investigation into Japanese Sect Activities in Western Australia," Australian Institute of Criminology, 2002, http://www.aic.gov.au/policing/case_studies/afp.html (accessed 2 June 2006).

42. Executive Office of the President, Executive Order 13228, *Establishing the Office of Homeland Security and the Homeland Security Council,* (Washington, DC: Executive Office of the President, 8 October 2001), http://frwebgate.access.gpo.gov/cgi-bin/getdoc.cgi?dbname=2001_register&docid=fr10oc01-144.pdf (accessed 2 June 2006).

43. Ibid.

44. Ibid.

45. Executive Office of the President, Executive Order 13231, *Critical Infrastructure Protection in the Information Age*, (Washington, DC: Executive Office of the President, 16 October 2001), http://frwebgate.access.gpo.gov/cgi-bin/getdoc.cgi?db name=2001_register&docid=fr18oc01-139.pdf (accessed 2 June 2006).

46. Moteff, "Critical Infrastructures," 10.
47. Office of Homeland Security, "National Strategy for Homeland Security," (Washington, DC: Office of Homeland Security, 16 July 2002), http://www.White House.gov/homeland/book/nat_strat_hls.pdf (accessed 2 June 2006).
48. Moteff, "Critical Infrastructures," 11.
49. Ibid.
50. President's Critical Infrastructure Protection Board, "National Strategy to Secure Cyberspace: Draft for Comment," (Washington, DC: President's Critical Infrastructure Protection Board, 18 September 2002), http://csrc.nist.gov/policies/cyberstrategy-draft.pdf (accessed 2 June 2006).
51. Bruce Berkowitz and Robert W. Hahn, "Cybersecurity—Who's Watching the Store?," *Issues in Science and Technology,* 19:3 (Spring 2006): 55–62, http://www.issues.org/19.3/berkowitz.htm (accessed 2 June 2006).
52. Moteff, "Critical Infrastructures," 11.
53. Ibid.
54. Department of Homeland Security, "History: Who Became Part of the Department of Homeland Security," http://www.dhs.gov/dhspublic/display?theme=10& content=5271 (accessed 2 June 2006).
55. U.S. Commission on National Security/Twenty-first Century, "Phase 3 Report," (Washington, DC: U.S. Commission on National Security/Twenty-first Century, 15 February 2001), 11–20, http://govinfo.library.unt.edu/nssg/PhaseIIIFR.pdf (accessed 2 June 2006); Moteff, "Critical Infrastructures," 8–9.
56. Executive Office of the President, "National Strategy to Secure Cyberspace," (Washington, DC: Executive Office of the President, February 2003), http://www.whitehouse.gov/pcipb/cyberspace_strategy.pdf (accessed 2 June 2006); Executive Office of the President, "National Strategy for the Physical Protection of Critical Infrastructures and Key Assets," (Washington, DC: Executive Office of the President, February 2003), http://www.whitehouse.gov/pcipb/physical_strategy.pdf (accessed 2 June 2006); Executive Office of the President, "National Strategy for Combating Terrorism," (Washington, DC: Executive Office of the President, February 2003), http://www.whitehouse.gov/news/releases/2003/02/counter_terrorism/counter_terrorism_strategy.pdf (accessed 2 June 2006).
57. Executive Office of the President, Executive Order 13286, *Amendment of Executive Orders, and Other Actions in Connection with the Transfer of Certain Functions to the Secretary of Homeland Security,* (Washington, DC: Executive Office of the President, 28 February 2003), see http://a257.g.akamaitech.net/7/257/2422/14mar20010800/edocket.access.gpo.gov/2003/pdf/03-5343.pdf (accessed 2 June 2006); Moteff, "Critical Infrastructures," 8–9.
58. Henry B. Hogue and Keith Bea, "Federal Emergency Management and Homeland Security Organization: Historical Developments and Legislative Options," (Washington, DC: Congressional Research Service, Report No. RL33369, 19 April 2006), 22–31, http://www.opencrs.com/rpts/RL33369_20060419.pdf (accessed 2 June 2006).
59. Department of Homeland Security (DHS), press conference on the Cyber Storm Cyber Security Preparedness Exercise, 10 February 2006, http://www.dhs.gov/dhspublic/display?content=5431 (accessed 2 June 2006).
60. Executive Office of the President, "Progress Report on the Global War on Terrorism," (Washington, DC: Executive Office of the President, September 2003): 5, http://www.White House.gov/homeland/progress/progress_report_0903.pdf (accessed 2 June 2006).
61. Ibid, 17–21.

62. Robert Cornell, "Australia's Approach to National Security", address at the Sydney Convention and Exhibition Centre, 28 April 2003, http://www.ag.gov.au/agd/WWW/ rwpattach.nsf/viewasattachmentpersonal/(CFD7369FCAE9B8F32F341D BE097801FF)~SEC+-+National+Security+Conference.pdf/$file/SEC+- +National+Security+Conference.pdf(accessed 2 June 2006).

63. Ibid.

64. Ibid.

65. Ibid.

66. Ibid.

67. National Counter-Terrorism Committee, *Critical Infrastructure Protection in Australia,* (Canberra: National Counter-Terrorism Committee, 25 March 2003), http:// www.tisn.gov.au/agd/WWW/rwpattach.nsf/VAP/(930C12A9101F61D43493D44C 70E84EAA)~NCTC+CIP+in+Australia+@+25+Mar+03+for+web+site.PDF/$file/ NCTC+CIP+in+Australia+@+25+Mar+03+for+web+site.PDF (accessed 2 June 2006).

68. Trusted Information Sharing Network, *Fact Sheet on the Trusted Information Sharing Network,* (Canberra: TISN, February 2006), http://www.tisn.gov.au/agd/WWW/ rwpattach.nsf/VAP/(930C12A9101F61D43493D44C70E84EAA)~ TISN+fact+sheet+Feb+06+final.pdf/$file/TISN+fact+sheet+Feb+06+final.pdf (accessed 2 June 2006).

69. Rachel Lebihan, "Government Belted on Security," *Australian Financial Review*, 26 June 2003, 19; James Riley, "Security Agency 'Needs Funds,'" *The Australian*, 19 August 2003, 26; Athol Yates, *Engineering a Safer Australia—Securing Critical Infrastructure and the Built Environment,* 51–68.

70. Department of Prime Minister and Cabinet, *Protecting Australia Against Terrorism,* (Canberra: Department of Prime Minister and Cabinet, 23 June 2004): 7–13, http:// www.dpmc.gov.au/publications/protecting_australia/index.htm#downloads (accessed 2 June 2006).

71. Council of Australian Government (COAG), *Communiqué of the COAG Meeting,* (Canberra: COAG, 5 April 2002), http://coag.gov.au/meetings/050402/ coag050402.pdf (accessed 2 June 2006).

72. National Counter-Terrorism Committee (NCTC), *National Counter-Terrorism Plan,* (Canberra: NCTC, June 2003): 8–9, http://www.nationalsecurity.gov.au/agd/www/ rwpattach.nsf/viewasattachmentPersonal/8142A6DCC44C2B44C A256E39000C7CD7/$file/NCT%20PLAN.pdf (accessed 2 June 2006). For the latest iteration of the National Counter-Terrorism Plan (dated September 2005), http:// www.nationalsecurity.gov.au/agd/WWW/rwpattach.nsf/VAP/(5738D F09EBC4B7EAE52BF217B46ED3DA)~NCTP_Sept_2005.pdf/$file/ NCTP_Sept_2005.pdf (accessed 2 June 2006).

73. NCTC, *National Counter-Terrorism Plan,* (June 2003 version): 7–8.

74. Peter Shergold, "Statement of the Secretary of the Department of Prime Minister and Cabinet—Organisational Restructure of the Department of Prime Minister and Cabinet," 23 May 2003, http://www.dpmc.gov.au/speeches/shergold/pmc_restructure_ 2003-05-23.cfm (accessed 2 June 2006).

75. Lincoln Wright, "No need to follow U.S. plan—Howard," *Canberra Times*, 24 May 2003, 4; Sandra Rossi, "Spooks Target IT Security," *Australian PC World*, 1 August 2003, 16.

76. John Howard, "Address to the Australian Defense Association," Melbourne, 25 October 2001, http://www.pm.gov.au/news/speeches/2001/speech1308.htm (accessed 2 June 2006).

77. Ibid.
78. Department of Foreign Affairs and Trade (DFAT), *Advancing the National Interest–Australia's Foreign and Trade Policy White Paper*, (Canberra: DFAT, 12 February 2003): 36–41, http://www.dfat.gov.au/ani/dfat_white_paper.pdf (accessed 2 June 2006); Department of Defense, *Australia's National Security: A Defense Update 2003*, (Canberra: Department of Defense, 26 February 2003): 13–14, http://www.defence.gov.au/ans2003/Report.pdf (accessed 2 June 2006). For a subsequent update of this paper, see DoD, *Australia's National Security: A Defense Update 2005*, (Canberra: DoD, 15 December 2005), http://www.defence.gov.au/update2005/defence_update_2005.pdf (accessed 2 June 2006).
79. Robert Cornall, "Australia's Approach to National Security," 9.
80. DFAT, *Transnational Terrorism: The Threat to Australia*, (Canberra: DFAT, 29 March 2004): 75—93, http://www.dfat.gov.au/publications/terrorism/transnational_terrorism.pdf (accessed 2 June 2006).
81. James A. Lewis, *Assessing the Risks of Cyber Terrorism*, 11.

Chapter 9: Speaking out of Both Sides of Your Mouth: Approaches to Perception Management in Washington, D.C. and Canberra

1. For a general history of perception management, see Philip M. Taylor, *Munitions of the Mind: A History of Propaganda from the Ancient World to the Present Era*, (Manchester: Manchester University Press, 2003).
2. Joseph S. Nye, Jr., *Soft Power: The Means to Success in World Politics*, (New York: Public Affairs, 2004), 22.
3. Ibid., 30–31.
4. Wilson P. Dizard, Jr., *Inventing Public Diplomacy: The Story of the U.S. Information Agency*, (Boulder, CO: Lynne Rienner Publishers, 2004), 17.
5. Ibid.
6. Ibid.
7. Ibid., 37–46.
8. National Security Council, NSC 4, *—Coordination of Foreign Information Measures*, (Washington, DC: NSC, 17 December 1947). http://www.fas.org/irp/offdocs/nsc-hst/nsc-4.htm (accessed 22 June 2006).
9. Dizard, Jr., *Inventing Public Diplomacy*, 45–47.
10. Ibid., 45.
11. Ibid., 63.
12. NSC, National Security Decision Directive 130, *—United States International Information Policy*, (Washington, DC: NSC, 6 March 1984). http://www.fas.org/irp/offdocs/nsdd/nsdd-130.htm (accessed 22 June 2006).
13. Daniel Kuehl, "The Information Component of Power and the National Security Strategy," in Alan Campden and Douglas Dearth, eds., *Cyberwar 3.0: Human Factors in Information Operations and Future Conflict*, (Fairfax, VA: AFCEA Press, 2000).
14. Executive Office of the President, Presidential Decision Directive 68, *—United States International Public Information*, (Washington, DC: Executive Office of the President, 30 April 1999), http://www.fas.org/irp/offdocs/pdd/pdd-68.htm (accessed 22 June 2006).
15. Ibid.; Leigh Armistead, "Fall from Glory: The Demise of the United States Information Agency during the Clinton Administration," *Journal of Information Warfare* 1:3, 107–124.

16. Dizard, Jr., *Inventing Public Diplomacy*, 220.

17. Godfrey Wiseman, "Australia's Information Service: How the Nation has Projected an Image Overseas," (unpublished dissertation, Royal Melbourne Institute of Technology, October 1990), 3.

18. Australian Information Service (AIS), *AIS Management Submission to the Price Waterhouse Review of AIS Operations*, (Canberra: AIS, July 1984), 2.1.

19. On the overall history of the Department of Information (DoI), see John Hilvert, *Blue Pencil Warriors: Censorship and Propaganda in World War II*, (St. Lucia, Queensland: University of Queensland Press, 1984); on the activities of FELO and other agencies, see Alan Powell, "The Human Element," in *War by Stealth: Australians and the Allied Intelligence Bureau, 1942–45*, (Carlton South, Victoria: Melbourne University Press, 1996).

20. Wiseman, "Australia's Information Service," 4.

21. Ibid.

22. Ibid., 2.

23. Geoffrey Forrester, "Policy Co-ordination in the Department of Foreign Affairs and Trade," in Patrick Weller, et al., *Reforming the Public Service*, (South Melbourne: Macmillan, 1993): 60–72.

24. Wiseman, "Australia's Information Service," 11.

25. Senate Standing Committee on Finance and Public Administration, *Management and Operation of the Department of Foreign Affairs and Trade*, (Canberra: Australian Government Publication Service, December 1992): 23–24.

26. Australian Journalists Association, *Submission to the Senate Standing Committee on Finance and Public Administration, Inquiry into the Department of Foreign Affairs and Trade*, (Canberra: Australian Journalists' Association, 30 November 1991), 11–23.

27. Mark Lever, "Foreign Information Service Axed," AAP News Wire, AAP496, 30 April 1996.

28. Ibid.

29. Innes Willox, "Recommendations Put Overseas Services Last," *The Age*, 25 January 1997, 7.

30. Denis Gastin, "False Economies Widen Rift," *The Australian*, 17 December 1996, 22.

31. Alison Broinowski, *About Face: Asian Accounts of Australia*, (Melbourne: Scribe Publications, 2003), 167–196.

32. Greg Sheriden, "Elite Unit to Fight Hanson in Asia," *The Australian*, 7 August 1997, 1.

33. Denis Gastin, "More Strategy, but No Action," *The Australian*, 11 August 1997, 11; Scott McKenize, "Hanson Hitman—Canberra Counts on the Power of One," *Daily Telegraph*, 22 August 1997, 9; Ian Stewart, "Asians Urge Canberra to Tackle Hanson Head-on," *The Australian*, 3 September 1997, 9.

34. Alison Broinowski, *About Face*, 183–84.

35. Ibid., 187–89.

36. Auditor-General, *Management of Australian Defense Force Deployments to East Timor, Audit Report No. 38, 2001–2002*, (Canberra: Australian National Audit Office, 2002), 110–111. http://www.anao.gov.au/website.nsf/publications/4a256ae90015f69bca256b810076f2f5/$file/38.pdf (accessed 22 June 2006).

37. Defense Science Board, *Report of the Defense Science Board Task Force on Strategic Communication*, (Washington, DC: Defense Science Board, September 2004), 26. http://www.acq.osd.mil/dsb/reports/2004-09-Strategic_Communication.pdf (accessed 22 June 2006).

38. Defense Science Board, *Report of the Defense Science Board Task Force on Managed Information Dissemination,* (Washington, DC: Defense Science Board, October 2001). http://www.acq.osd.mil/dsb/reports/mid.pdf (accessed 22 June 2006).

39. Ibid., 59.

40. Franklin Foer, "Flacks Americana: John Rendon's Shallow PR War on Terrorism," *New Republic* 226: 19, 20 May 2002, 24.

41. Defense Science Board, *Report of the Defense Science Board Task Force on Strategic Communication,* (Washington, DC: Defense Science Board, September 2004), 24.

42. Ibid.

43. Ibid., 21.

44. Ibid., 25.

45. Ibid.

46. Ibid.

47. Department of State, "President and Secretary Honor Ambassador Karen Hughes at Swearing-In Ceremony," 9 September 2005, http://www.state.gov/secretary/rm/2005/52846.htm (accessed 22 June 2006).

48. United States Advisory Commission on Public Diplomacy, *2005 Report,* (Washington, DC: United States Advisory Commission on Public Diplomacy, 7 November 2005). 8. http://www.state.gov/documents/organization/55989.pdf (accessed 22 June 2006).

49. Pew Research Center, *2006 Pew Global Attitudes Report,* (Washington, DC: Pew Research Center, 13 June 2006), 8–14. http://pewglobal.org/reports/pdf/252.pdf (accessed 22 June 2006).

50. Alexander Downer, "Media Release: Australian Television Service to the Asia-Pacific Region," (Canberra: Department of Foreign Affairs and Trade, 20 June 2001), http://www.dfat.gov.au/media/releases/foreign/2001/fa084_01.html (accessed 22 June 2006).

51. Alexander Downer, "Media Release: Australian Television Service in the Asia-Pacific Region," (Canberra: Department of Foreign Affairs and Trade, 19 March 2001), http://www.dfat.gov.au/media/releases/foreign/2001/fa038_01.html (accessed 22 June 2006).

52. Alison Broinowski, "Bali as Blowback: Australia's Reputation in Asian Countries," *Sydney Papers*, Autumn 2003, 66–67.

53. Department of Foreign Affairs and Trade, *Transnational Terrorism: The Threat to Australia*, (Canberra: Department of Foreign Affairs and Trade, 29 March 2004), 66, http://www.dfat.gov.au/publications/terrorism/transnational_terrorism.pdf (accessed 22 June 2006).

54. Ibid.

55. Ibid., 67.

56. Broinowski, "Bali as Blowback," 66–67.

57. Department of Foreign Affairs and Trade, *Advancing the National Interest*, (Canberra: Department of Foreign Affairs and Trade, 12 February 2003), 124—131, http://www.dfat.gov.au/ani/dfat_white_paper.pdf (accessed 22 June 2006).

58. Department of Foreign Affairs and Trade, *In the National Interest*, (Canberra: Department of Foreign Affairs and Trade, August 1997), 78, http://www.dfat.gov.au/ini/whitepaper.pdf (accessed 22 June 2006).

59. Department of Foreign Affairs and Trade, *Transnational Terrorism: The Threat to Australia*, 104–105.

60. Carl Oatley, "Australia's National Security: A Look to the Future," (Canberra: Australian Defence Studies Centre, Working Paper No. 61, October 2000), http://pandora.nla.gov.au/pan/33932/20030320/idun.itsc.adfa.edu.au/ADSC/national_secu.ty_-_pagemake.pdf (accessed 22 June 2006).

61. On organizational coordination, see the previous chapter in this volume, "A Tale of Two Cities," X1–X2.

62. John Fitzgerald, "Who Cares What They Think? John Winston Howard, William Morris Hughs and the Pragmatic Vision of Australian National Security Policy," in Alison Broinowski, ed., *Double Vision: Asian Accounts of Australia*, (Canberra: Pandanus Press, 2004), 19.

63. On Australian organizational informality, see the previous chapter in this volume, "A Tale of Two Cities," X1–X2.

64. For the United States, see "The White House, National Strategy for Combating Terrorism," (Washington, DC: The White House, February 2003), 19–24, http://www.whitehouse.gov/news/releases/2003/02/counter_terrorism/counter_terrorism_strategy.pdf (accessed 22 June 2006). For Australia, see Department of Foreign Affairs and Trade, *Transnational Terrorism: The Threat to Australia,* 76–98.

65. Nye, Jr., *Soft Power*, 110.

66. Susan B. Epstein, "U.S. Public Diplomacy: Background and the 9/11 Commission Recommendations," (Washington, DC: Congressional Research Service, Report No. RF32607, 1 May 2006), 12, http://fpc.state.gov/documents/organization/66505.pdf (accessed 22 June 2006).

67. Nye, Jr., *Soft Power*, 111.

68. Susan B. Epstein, "U.S. Public Diplomacy" 12.

Chapter 10: Conclusion

1. Arquilla, John, and David Ronfeldt, The Emergence of Noopolitik: Toward an American Information Strategy (Washington, DC: Rand, 1999)

2. T Toffler, Alvin, *Future Shock,* (New York: Bantam, 1970).

Index

About the Authors

LEIGH ARMISTEAD

Currently the strategic information assurance development manager for Honeywell Technology Solutions Inc, Leigh was also the editor of *Information Operations: Warfare and The Hard Reality of Soft Power*. A retired U.S. Naval Officer and Master Faculty of IO at the Joint Forces Staff College, he is currently enrolled in a PhD program at Edith Cowan University in Perth, Australia. Leigh has published a number of articles on IO in addition to chairing numerous professional IO conferences around the world, including the International Conference on Information Warfare in 2006 and the IQPC IO Conference in the United Kingdom from 2002 to 2005. Selected five years in a row as a research fellow for the International National Security Studies program to conduct IO-related research, he also helped to develop an online IO course for the National Security Agency.

MAURA CONWAY

Maura is a lecturer in the School of Law and Government at Dublin City University where she teaches in the MA programs in International Relations and International Security and Conflict Studies. Previously, she was a teaching fellow in the School of International Relations at the University of St. Andrews, Scotland and was awarded her PhD from the Department of Political Science at Trinity College Dublin, Ireland. Her research interests are in the area of terrorism and the Internet. She is particularly interested in cyberterrorism and its portrayal in the media, and the functioning and effectiveness of terrorist websites. Along with a number of book chapters, Maura has also been published in *First Monday*, *Current History*, the *Journal of Information Warfare*, and elsewhere.

ZACHARY P. HUBBARD

A retired U.S. Army officer, Zach was the head of the Information Warfare Division of the Joint Forces Staff College from April 1998 to April 2001. While there, he authored the first edition of the *Joint Information Operations Planning Handbook*, which became the basis for the Department of Defense's Joint IO Planning Course. His twenty-four year military career included service in the U.S., Europe, Africa, and the Middle East. After retiring from the Army in 2001, Zachary joined the defense industry as a corporate division manager specializing in IO. He simultaneously served as a program manager in the DARPA. Cyber Panel Project. Following the successful completion of the project, Zachary became a systems engineering and technical assistance representative for the air force, developing IO requirements and capabilities for future air force C2 and ISR systems. He has lectured and written extensively on IO topics. Zachary currently resides in Johnstown, Pennsylvania.

Daniel Kuehl

Dr. Daniel Kuehl is the senior information operation instructor at the National Defense University, in Washington, D.C., where he has taught for the last fifteen years. Recognized throughout the Department of Defense and academia as one of leading lecturers in the area of Information Operations, he is often requested to speak as well as chair conferences throughout the world. In addition, Dr. Kuehl has authored a number of articles and thought pieces in journals on the perception management portions of IO, as well as the need for changes in the training and the standardization of IO education in this critical warfare area.

Jeff Malone

A retired Australian Army officer, Jeff has been closely involved in the development of IO within the Australian Defense Force and he has been published and presented extensively on IO internationally. In 2003, Jeff was appointed the chief of defense force fellow on the basis of his work in the IO field. He is currently enrolled in a PhD program at the University of New South Wales, the Australian Defense Force Academy, in Canberra, and is an academic member of the Research Network for a Secure Australia, an Australian research consortium focusing on critical infrastructure protection and aligned issues. Jeff is currently employed by Noetic Solutions Pty. Ltd. as a consultant/analyst.

Corey Schou

Dr. Corey Schou is a professor of Infomatics at Idaho State University in Pocatello, Idaho and is a founding member of Colloquium of Information System Security Educators, which is a group that brings together industry, academia, and government to understand the needs of education in information assurance in the United States. In this role, he serves as the director of the National Information Assurance Training and Education Program, which helps to build standards and curriculum for IA that is certified and accredited. A long-time advocate of advancing the state of education in the IA world, he has worked closely with the National Security Agency in hosting the Committee for National Security Standards annual reviews.

Pascale Combelles Siegel

Mrs. Combelles Siegel is president of Insight Through Analysis, an independent consultancy company based in McLean, VA. She has conducted numerous post-conflict analyses, including a review of the Pentagon's embedding policy and information operations during Operation Iraqi Freedom in 2004, and an analysis of American sensitivities toward military casualties in 2002. She also conducted lessons-learned analyses of the media war during Operation Allied Force in 1999 and of NATO's information campaign in Bosnia from 1996 to 1997). Major publications include: "Lessons from the Pentagon's Embedding Program," "Information Operations During Operation Iraqi Freedom," "The Myth of the Zero Casualties Syndrome," "Information Campaigns in Peace Operations" (co-author), and "Target Bosnia: Integrating Information Activities in Peace Operations: The NATO-led Operations in Bosnia-Herzegovina: December 1995-1997."

David Wolfe

A retired U.S. Air Force officer, Dave completed his twenty-eight-year career as a faculty instructor in the Joint Command, Control, and Information Warfare School of the Joint Forces Staff College from September 2000 to September 2002. While there, he was requested by name following the events of 9/11 and augmented the initial cadre of officers who formed the Directorate of Homeland Security within the United States Joint Forces Command. Prior to this, he served with the U.S. Joint Forces Command as a member of the Joint Warfighting Center's Deployable Training Team. As an observer-trainer on that team, he instructed Joint Task Force (JTF) staffs from all services during military exercises around the globe, both ashore and afloat, and deployed to assist combatant commanders during contingency operations and conflicts. After retiring from the Air Force in 2002, David joined the defense industry, where he is employed by Honeywell as Team Lead, Naval Network Warfare Command.